Crystal Structure Determination

William Clegg

Professor of Structural Crystallography
at the University of Newcastle upon Tyne,
with a Joint Appointment
at CCLRC Daresbury Laboratory

Series sponsor: **ZENECA**

ZENECA is a major international company active in four main areas of business:
Pharmaceuticals, Agrochemicals and Seeds, Specialty Chemicals, and Biological Products.

ZENECA's skill and innovative ideas in organic chemistry and bioscience create products
and services which improve the world's health, nutrition, environment, and quality of life.
ZENECA is commited to the support of education in chemistry and chemical engineering.

OXFORD   NEW YORK   TOKYO
OXFORD UNIVERSITY PRESS
1998

Oxford University Press, Great Clarendon Street, Oxford OX2 6DP

Oxford   New York
Athens   Auckland   Bangkok   Bogota   Bombay
Buenos Aires   Calcutta   Cape Town   Dar es Salaam
Delhi   Florence   Hong Kong   Istanbul   Karachi
Kuala Lumpur   Madras   Madrid   Melbourne
Mexico City   Nairobi   Paris   Singapore
Taipei   Tokyo   Toronto   Warsaw

and associated companies in
Berlin   Ibadan

Oxford is a trade mark of Oxford University Press

Published in the United States
by Oxford University Press Inc., New York

A catalogue record for this book is available from the British Library

Library of Congress Cataloging in Publication Data

Clegg, William, Prof.
Crystal structure determination / William Clegg.
(Oxford chemistry primers; 60)
Includes index.
1. X-ray crystallography. I. Title. II. Series.
QD945.C56   1998   548'.83—dc21   97-46780

ISBN 0 19 855901 1

Typeset by
EXPO Holdings, Malaysia

Printed in Great Britain by
The Bath Press, Avon

Series Editor's Foreword

Single-crystal X-ray diffraction is the bedrock structural technique in modern day chemistry. The majority of inorganic, organometallic and organic molecular and lattice geometries have been established in this way. So it is important for all students to have an understanding of the basis, strengths, precision and limitations of this technique.

Oxford Chemistry Primers are designed to give a concise introduction to all chemistry students by providing the material that would normally be covered in an 8–10 lecture course. As well as giving up-to-date information, this series provides explanations and rationales that form the framework of understanding inorganic chemistry. Bill Clegg here gives us an approachable way to grasp the basis and application of X-ray diffraction. The excellent illustrations allow one to follow the origin of diffraction patterns, structure solution methods, three-dimensional structures and forms of disorder. He also gives a glimpse of the exciting new opportunities attainable with the intense X-ray sources.

John Evans
Department of Chemistry, University of Southampton

Preface

Although there are numerous excellent accounts of X-ray crystallography, I have always had difficulty in recommending one for undergraduate chemistry students, for most of whom the subject is treated in a short course of lectures, perhaps with an associated calculation class or practical experiment. The available texts are simply too long and detailed. The Oxford Chemistry Primer format and size are ideal for this purpose. This book and my own second-year course in Newcastle have evolved together over the last couple of years, and I hope it will provide valuable support material for such courses elsewhere.

The Primer length restriction is also a tremendous challenge. I have tried to concentrate on the significance and importance of crystal structure determination in the context of modern chemistry, and use illustrations and examples in preference to rigorous mathematical treatment; some purists will not approve! Many details have had to be left out, and the book is certainly not a complete practical manual.

I am very grateful to John Evans for his patience in the long period since this Primer was first suggested, and for his helpful suggestions for improvement to the first draft. I would also like to thank Marjorie Harding, Paul Raithby, Mark Elsegood and Andy Edwards for their comments and corrections. The responsibility for any remaining errors and shortcomings is, of course, entirely my own.

Newcastle upon Tyne   Bill Clegg
December 1997

Contents

1 The basis of the method

1.1 Introduction

A knowledge of the structure of both molecular and non-molecular materials is one of the fundamental aims of chemistry and is essential for a proper understanding of the physical and chemical properties of the materials. The term 'structure' has many meanings; here we take it to be the relative positions of the atoms or ions which make up the substance under study and hence a geometrical description in terms of bond lengths and angles, torsion angles, non-bonded distances and other quantities of interest. This knowledge makes possible the pictorial representation of chemical structures which are to be found throughout the literature of teaching and research in chemistry; typical examples are shown in Fig. 1.1. Knowledge of the structure may be sought

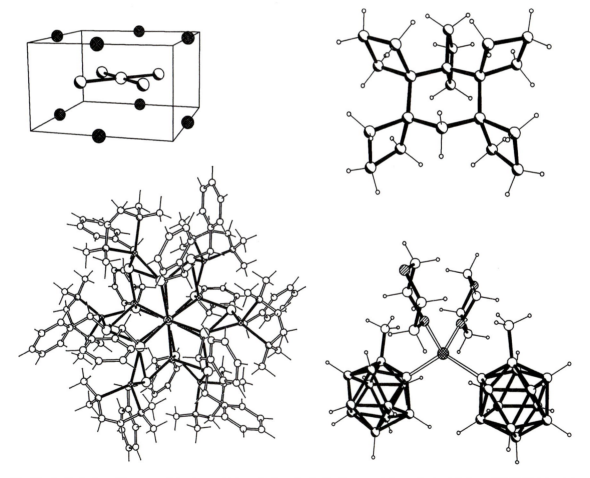

Fig. 1.1 An illustration of the range of structures which can be investigated by X-ray crystallography: top left, the salt $K_2[PdCl_4]$; top right, a relatively small organic molecule; bottom right, an organometallic complex; bottom left, a polynuclear metal cluster complex.

simply as a means of identifying a newly synthesized compound and understanding how it was formed, or the detailed geometry may be important for further investigations of reactivity, bonding, structure–energy relationships, etc.

Many experimental methods of probing the structure of a material are based on its absorption or emission of radiation; these are various forms of spectroscopy. Absorption takes place when the frequency ν of the radiation, and hence its quantum energy $h\nu$, matches a difference in certain energy levels in the sample. The observed frequencies of absorption thus provide information about energy levels and, from this, something can be deduced about the structure of the material, based on a substantial body of accumulated experience. For example, chemical shifts and coupling constants in proton NMR spectroscopy, obtained from the measured absorption spectrum, may indicate the number, chemical type, and relative proximity of the hydrogen atoms in a molecule and thus provide information on the connectivity (which atoms are bonded together); by more detailed NMR experiments some interatomic distances can be calculated.

In most spectroscopic techniques, what is measured is the variation of intensity of radiation passing through the sample as its frequency (or wavelength) is varied, in a particular direction; the intensity variation is produced by absorption of particular frequencies, leading to energy changes in the sample. This Primer is concerned with some diffraction methods, which are based on a different interaction of radiation with matter, usually in the solid state. Here we usually keep the wavelength fixed and measure the variation of intensity with direction, i.e. the scattering of *monochromatic* radiation is measured. From these measurements it is possible to work out the positions of the atoms in the sample and hence obtain a complete geometrical description of the structure. The intensity variation is caused by interference effects, also known as *diffraction*.

These methods are, therefore, capable of providing much more detailed structural information than spectroscopic methods. There are, however, limitations on the types of materials which can be studied, as we shall see.

A substantial body of rather complex mathematics forms the theory of diffraction methods for crystal structure determination. Fortunately it is not necessary to master this in order to understand the principles and application of the subject. This is, to a large extent, true even for those who carry out research in crystallography, because almost all the calculations are usually performed by computers. Advances in the subject since its birth near the beginning of the twentieth century have very much been in parallel with developments in computing, and modern personal computers are of sufficient power to make many crystal structure determinations very much faster than the days or weeks suggested by even quite recent textbooks. Some of the other modern technological developments which have contributed to this dramatic increase in speed are described later in this book. The level of mathematics has deliberately been kept relatively low in this treatment of the subject. The fundamental equations for the diffraction process are given, but they are also illustrated with analogies in words and diagrams in order to clarify their meaning.

Crystal structure determination can be applied to a wide range of size of structures, from very small molecules and simple salts to synthetic and natural

Monochromatic, literally 'single coloured', means having a single wavelength.

Diffraction effects are a characteristic behaviour of waves, including X-rays, light and other forms of electromagnetic radiation. An example can easily be seen by looking at a yellow street lamp (monochromatic light) through a finely woven fabric such as an umbrella.

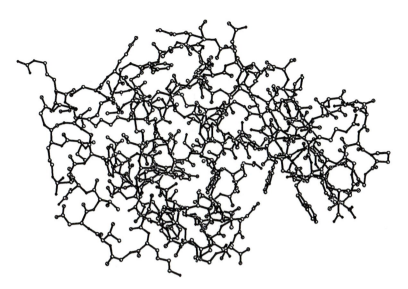

Fig. 1.2 The structure of lysozyme, a relatively small protein; hydrogen atoms are omitted. The molecule contains C, N, O and S atoms.

polymers and to biological macromolecules such as proteins (Fig. 1.2) and even viruses. This Primer is concerned mainly with chemical applications, but some indication is given of the differences encountered when working with larger scale biological systems.

1.2 The eye and microscope analogy

Objects of macroscopic size are visible to us because they scatter light falling on them. Our eyes intercept some of the scattered light and the function of the eye lens is to bring together the bundle of light rays, recombining the individual rays into an image on the retina (Fig. 1.4). Light consists of waves, and each scattered light ray has a particular intensity and a particular phase relative to other scattered rays, resulting from the scattering which produced it (Fig. 1.3). These relative intensities and phases, in turn, determine the nature of the image formed in the eye, which is understood by the brain as a representation of the object being viewed. Thus information on the shape (structure) of the object is carried in the intensities and phases of the light waves scattered by it.

Since objects with different shapes can be distinguished by looking at them, it follows that they must have different, individual scattering patterns.

For smaller objects the eye needs help from more powerful lenses which can produce a larger image. The principle of operation is just the same: a proportion of the scattered light is collected by the microscope lens system and is *refracted* (the directions of the light rays are changed) to bring it all back together (the individual rays are combined by addition, with the correct relative amplitudes and phases) in the observer's eye (Fig. 1.4). Note that it is not necessary to capture all the scattered light, but the observed image becomes less clear as the proportion of light collected is reduced: a good quality image is produced by a wide objective lens close to the object.

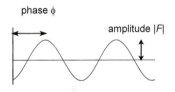

Fig. 1.3 Amplitude and phase of a wave.

Visible light also has a range of wavelengths λ and frequencies v, such that $\lambda v = c$, the velocity of light; for simplicity here and for comparison with the X-ray diffraction experiment, we consider light of just one wavelength, i.e. monochromatic light.

Refraction is the alteration in the direction of travel of light as it passes from one medium into another with a different *refractive index*. It is responsible, for example, for the apparent bending of a drinking straw in a glass of water, because water and air have quite different refractive indices. Refraction should not be confused with diffraction, a quite different phenomenon despite the similar name.

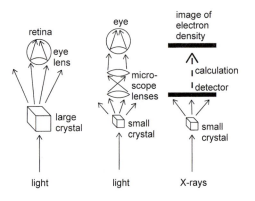

Fig. 1.4 Left and centre, the function of the eye and microscope lenses for the recombination of scattered light; right, the equivalent two-stage process in X-ray diffraction.

The *angstrom unit*, Å is not strictly permitted by the SI rules, but is widely used in structural chemistry because of its convenient size. 1 Å $= 100$ pm $= 0.1$ nm.

The lower limit on the size of objects which can be clearly seen with a microscope of sufficient magnifying power is set, not by optical engineering capabilities, but by the wavelength of light itself (in the range 400–700 nm). Objects which are much smaller than this, such as individual molecules (which are 100–1000 times smaller), do not give any significant scattering of the light. In order to 'see' the structure of molecules, it is necessary to resolve the component atoms, which are of the order of one or a few *angstroms* in size. Instead of visible light, this means using X-rays. So, in principle, what we need is an X-ray microscope in order to observe molecular structure.

Focusing of extremely high intensity X-rays has recently been achieved using special methods, but they are not of general application.

Quite apart from safety issues and the need to provide suitably sensitive detectors to record the X-rays, such an instrument does not exist, because conventional lenses cannot be used to focus X-rays. The scattering of X-rays by molecules does, indeed, occur, but the scattered X-rays cannot be physically recombined to form an image.

Note that it is perfectly possible to record the pattern of light scattered by an object in an analogous way; some examples will be used in Section 1.4. Although this is not commonly done, a variant of it is a valuable procedure used by mineralogists, in identifying and characterizing mineral specimens.

The situation is, however, not hopeless, because the pattern of scattered X-rays can be directly recorded, either on photographic film or on a variety of other X-ray sensitive detectors, then the recombination which is impossible physically can be performed mathematically, with the aid of computers: the mathematics involved is well established, but it requires a considerable amount of calculation. Thus the experiment to determine a molecular structure falls into two parts, recording the X-ray scattering pattern and carrying out the recombination subsequently by mathematics (Fig. 1.4), and is no longer instantaneous like viewing an object through a microscope.

The technique is known as crystal structure determination because the object studied is actually a small crystalline sample rather than a single molecule, which it would be impossible to hold in the X-ray beam for the duration of the experiment and which would give an immeasurably weak scattering pattern. In a crystal, there are large numbers of identical molecules, locked in position in a regular arrangement, which together give significant scattering. The method can only be used for samples which can be obtained in a suitable crystalline form.

When the method is successful, it provides an image of the molecular structure. More precisely, it locates the components of the material which interact with the incident X-rays and scatter them. These are the electrons in the atoms. Although each individual electron/X-ray interaction is instant-

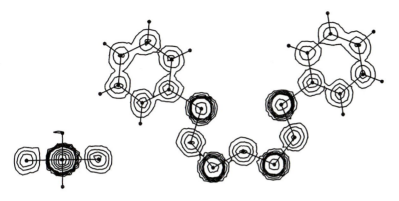

Fig. 1.5 An electron density map, with the positions of atoms and bonds marked. Note that the rather wiggly contours here and in other figures are an artefact of the computer program used.

aneous, the total time taken to record the scattering pattern is usually hours or days, and even the most rapid methods available are very slow compared with the movement of electrons, so the picture that results is of a time-averaged electron density (Fig. 1.5). Concentrations of electron density in the image correspond to atoms, somewhat spread out by time-averaged vibration, and the results are usually interpreted and presented as atomic positions, but there are some important consequences of the fact that the primary result is the location of electron density, which will be discussed later.

One very important consequence of the need to divide the overall experiment into two parts instead of directly recombining the scattered X-rays to generate an image is that some of the information in the scattered X-rays is lost. When the X-ray scattering pattern is recorded, the individual wave amplitudes are retained as relative intensities (intensity is proportional to the square of amplitude), but the relative phases are lost. This makes the mathematical reconstruction stage much less straightforward. It is one of the fundamental challenges of crystallography and methods of dealing with 'the phase problem' have been major research projects throughout the history of the subject.

$I \propto |F|^2$

1.3 Fundamentals of the crystalline state

A perfect crystalline solid material is made up of a large number of identical molecules which are arranged in a precisely regular way repeated in all directions, to give a highly ordered structure. Even for a microscopic crystal, the repetition is effectively infinite on an atomic scale. This repetition of a structural unit by translation, to form a space-filling, three-dimensional crystal, is a type of symmetry, which occurs in all crystalline solids, whether or not they also show other forms of symmetry such as rotation or reflection.

In two dimensions, such *translation* symmetry is familiar in the form of patterns on wallpapers, flooring and other manufactured materials. A two-dimensional projection of part of a real crystal structure is shown in Fig. 1.6(a). The basic structural unit is a single molecule. All the molecules are identical and repetition by translation gives the complete two-dimensional pattern.

The basic unit of a crystal structure may not be a single molecule, but a number of ions, an assembly of a few molecules, or other unit, which is then repeated. Real structures also show various kinds of defects and irregularities.

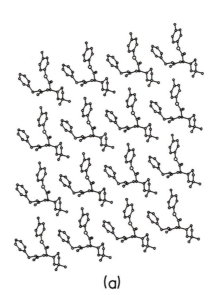

(a)

(b)

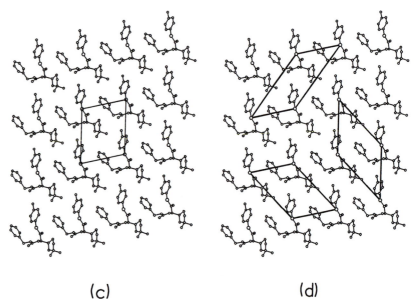

(c)

(d)

Fig. 1.6   (a) A two-dimensional projection of an organic crystal structure, showing 16 identical molecules; (b) the two-dimensional *lattice* for this pattern; (c) a *unit cell* for the pattern; (d) other possible choices of unit cell for the same pattern.

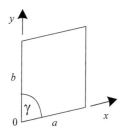

Fig. 1.7   The geometry of the unit cell for Fig. 1.6.

If each molecule is represented by just a single point (placed, for example, on the same atom in each molecule), the result is a regular array of points, which shows the repeating nature of the structure but not the actual form of the basic structural unit (Fig. 1.6(b)). This array of identical points, equivalent to each other by translation symmetry, is called the *lattice* of the structure.

To define the repeat geometry of the structure, a parallelogram of four lattice points is chosen, and is called the *unit cell* of the structure; it has two different sides and one included angle (Figs 1.6(c) and 1.7). Obviously, many different choices of unit cell are possible for any one lattice (Fig. 1.6(d)); there

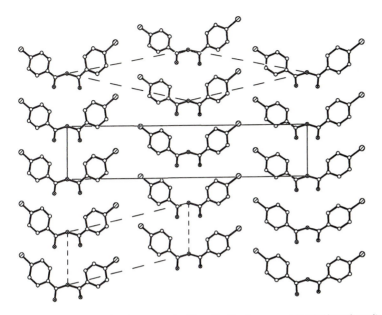

Fig. 1.8 A structure with reflection symmetry. The reflection lines run vertically along the unit cell edges and through the cell centre, bisecting each molecule. There are also glide reflection lines half-way between these (see later). The conventional centred rectangular unit cell is outlined, together with two possible primitive cells, each of half the area.

are conventions to guide this choice. In the absence of any rotation or reflection symmetry in the structure, the conventional unit cell has sides as short as possible, $a \leq b$, and an angle as close as possible to 90°. Inspection shows that each unit cell in Fig. 1.6(c) contains parts of several molecules, and the total contents are just one molecule. Each unit cell contains the equivalent of one *lattice point* (one repeat unit).

The presence of rotation or reflection symmetry in a crystal structure, relating molecules or parts of molecules to each other, imposes restrictions on the geometry of the lattice and unit cell. For example, four-fold rotation symmetry in a two-dimensional lattice requires a square unit cell with two equal sides and a 90° angle, while reflection symmetry gives a 90° angle but still allows the two unit cell sides to be of different length (Fig. 1.8).

In three dimensions, a unit cell has three sides and three angles (Fig. 1.9). Conventionally, the three lengths are called a, b, c, and the angles α, β, γ, such that α lies between the b and c axes, i.e. opposite a. In the absence of any rotation or reflection symmetry, the three axes have different lengths and the three angles are different from each other and from 90°. Rotation and reflection symmetry impose restrictions and special values on the unit cell parameters. On the basis of these restrictions, crystal symmetry is broadly divided into seven types, called the seven *crystal systems*. Table 1.1 shows their names, minimum symmetry characteristics and unit cell geometries. Note that inversion symmetry does not impose any lattice geometry restrictions, since every three-dimensional lattice has inversion symmetry anyway, whether the complete structure represented by the lattice is centrosymmetric or not. As we shall see later, this has important consequences in X-ray diffraction.

Individual objects, such as molecules, can display rotation symmetry of any order C_2, C_3, ... up to C_∞ rotation axes, but only C_2, C_3, C_4 and C_6 axes can

For an introduction to symmetry in chemistry, reference should be made to standard chemistry text books.

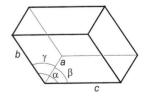

Fig. 1.9 A three-dimensional unit cell.

Table 1.1 Crystal systems

Crystal system	Essential symmetry	Restrictions on unit cell
Triclinic	none	none
Monoclinic	one two-fold rotation and/or mirror	$\alpha = \gamma = 90°$
Orthorhombic	three two-fold rotations and/or mirrors	$\alpha = \beta = \gamma = 90°$
Tetragonal	one four-fold rotation	$a = b; \alpha = \beta = \gamma = 90°$
Rhombohedral	one three-fold rotation	$a = b = c; \alpha = \beta = \gamma \, (\neq 90°)$
Hexagonal	one six-fold rotation	$a = b; \alpha = \beta = 90°; \gamma = 120°$
Cubic	four three-fold rotation axes	$a = b = c; \alpha = \beta = \gamma = 90°$

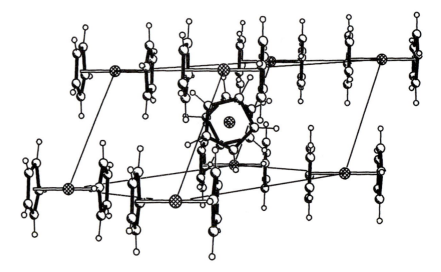

Fig. 1.10 The structure of one crystalline form of ferrocene, showing that the surroundings of each molecule do not have five-fold rotation symmetry.

be found in crystals. This does not mean that molecules with C_5 symmetry, for example [such as ferrocene $(C_5H_5)_2Fe$, or buckminsterfullerene C_{60}] cannot form crystalline solids, but the rotation symmetry does not apply to the surroundings of the molecule and to the structure as a whole (Fig. 1.10).

On the other hand, crystals can have other types of symmetry element not possible in single finite molecules, in which rotation or reflection is combined with translation to give, respectively, *screw axes* and *glide planes*. Figure 1.11 illustrates the difference between simple reflection and glide reflection in a two-dimensional pattern. Reflection symmetry is familiar in everyday life; the two mirror-related objects lie directly opposite each other, reflected from each other across the mirror line (in two dimensions) or mirror plane (in three dimensions). Glide reflection involves displacement of the two mirror images relative to each other by exactly half of a repeat unit of the pattern. In a two-dimensional pattern such as in Fig. 1.11, there is only one possibility for the direction of glide, parallel to the glide line. In a three-dimensional crystal structure, the direction of glide can, in most cases, be parallel to either of two different axes or along the diagonal between them. Similarly, screw axes combine a simple rotation with a translation along the direction of the axis.

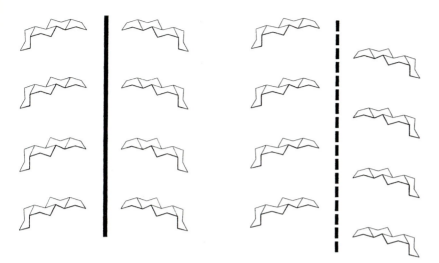

Fig. 1.11 A two-dimensional illustration of simple reflection (left) and glide reflection (right) for a regularly spaced repeated motif. Because glide reflection provides more efficient packing of molecules than does simple reflection, space groups with glide planes tend to be observed more often than those with mirror planes. A similar argument applies to screw axes versus simple rotation axes.

An example of glide reflection symmetry occurring in the same structure as simple reflection symmetry can be seen in Fig. 1.8.

For some structures with more than purely translation symmetry, it is convenient and conventional to choose a unit cell which contains more than one lattice point, because the resultant unit cell geometry displays the symmetry more clearly. The unit cell in Fig. 1.8 has a lattice point at each corner, but also an entirely equivalent lattice point at its centre; this *centred* unit cell is a rectangle, with a 90° angle, whereas a *primitive* unit cell with lattice points only at its corners and with half the area, would not be. A three-dimensional example is the *face-centred* cubic structure of many metals, in which the centre of each face of the unit cell is equivalent to all the cell corners (Fig. 1.12). This unit cell has four times the volume of the smallest possible primitive unit cell, but has the advantage of displaying the cubic symmetry clearly.

Symmetry elements in a single molecule all pass through one point, and the various possible combinations of symmetry elements are known as

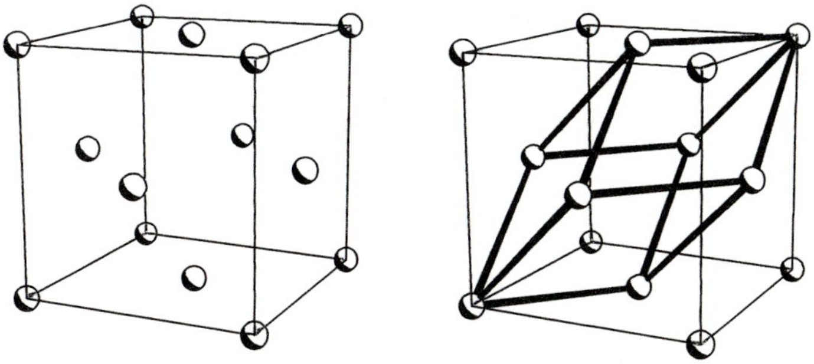

A *primitive* unit cell has lattice points only at its eight corners; all of these, by definition of a lattice, are entirely equivalent. A *centred* unit cell has other, also entirely equivalent, lattice points. For a two-dimensional lattice there is only one form of centring, with an equivalent lattice point exactly at the centre of each unit cell. For a three-dimensional lattice, various different types of centring are possible, with lattice points at the centres of pairs of opposite faces, at the centres of all faces, or at the very centre (the body centre) of each unit cell.

Fig. 1.12 The face-centred cubic structure of many metals; left, the conventional cubic unit cell; right, a primitive unit cell with one-quarter the volume, having lattice points only at its corners.

A *point group* is the complete collection of all symmetry elements passing through a central point, describing the symmetry of an individual object. A *space group* is the complete collection of all symmetry elements for an infinitely repeating pattern. Both can be elegantly treated by mathematical group theory.

The *International Tables for Crystallography* are published for the International Union of Crystallography by Kluwer Academic Publishers, Dordrecht, The Netherlands.

A *proper rotation* is a simple rotation axis. An *improper rotation* combines the operation of rotation with either inversion (crystallography) or reflection in a perpendicular plane (spectroscopy).

point groups. In a crystal, symmetry elements do not all pass through one point, but they are regularly arranged in space in accordance with the lattice translation symmetry. There are exactly 230 possible arrangements of symmetry elements in the solid state; these are called the 230 *space groups*, and their symmetry properties are well established and available in standard reference books and tables, the most comprehensive and widely used being the *International Tables for Crystallography*, Volume A.

Some popular misconceptions about lattices and unit cells should be dispelled. The term 'lattice' is often used as a synonym for 'structure', but this is incorrect, because the lattice shows only the repeating nature of the structure, not the detailed contents. *Any* point in a crystal structure can be chosen as a lattice point and the lattice constructed from all the equivalent points (with identical environments in identical orientation); conventions apply to the choice of lattice points relative to the positions of symmetry elements in the structure (in particular, it is conventional to place lattice points on inversion centres when they are present), and it is the exception rather than the rule for an atom to lie on a lattice point in any other than simple high-symmetry structures. The unit cell is the repeat unit (building block) of any crystal structure and so contains a small whole number of molecules, but in most structures the molecules lie across unit cell edges and faces rather than being neatly contained within these purely mathematical constructions (see, for example, Figs 1.6 and 1.8).

Pure lattice translation symmetry relates individual unit cells to each other. If a structure has any other symmetry as well, then this symmetry relates atoms and molecules within one unit cell to each other. Thus the unique, independent part of the structure is usually only a fraction of the unit cell, the fraction depending on the amount of symmetry present. This unique portion is called the *asymmetric unit* of the structure. Operation of all the rotation, reflection, inversion and translation symmetry elements of the space group on this asymmetric unit generates the complete crystal structure. The asymmetric unit may consist of one molecule, more than one molecule, or a fraction of a molecule, which itself possesses symmetry.

One further point should be made about symmetry. Different symbols are used for symmetry elements, and for combinations of them (point groups and space groups) in different fields of science. For rotation, reflection and inversion symmetry elements, which can occur both in individual molecules and in solid-state structures, the correspondence of the two sets of symbols is shown in Table 1.2. There are good reasons for the differences in the symbols, and also for the different definitions of the so-called '*improper rotations*', but they do lead to confusion. The Schoenflies notation, used in molecular spectroscopy, produces convenient and compact symbols for point groups, while the Hermann–Maugin notation, used in crystallography, is much better suited to space group representation, some of which will be seen later in examples. The details of symbols for glide planes and screw axes are beyond the scope of this short text.

1.4   Diffraction of X-rays by molecules and crystals

Figure 1.13 shows part of the pattern of scattered X-rays produced by a single crystal. The complete pattern can only be recorded by rotating the crystal in

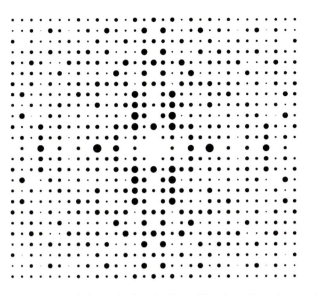

Fig. 1.13   A computer-generated reproduction of an X-ray diffraction pattern of a type obtained by one form of X-ray camera.

Table 1.2   Correspondence of symbols for symmetry elements in the Hermann–Maugin (crystallography) and Schoenflies (spectroscopy) systems of notation.

Crystallography	Spectroscopy
Proper rotations	
2	C_2
3	C_3
4	C_4
6	C_6
Improper rotations	
$\bar{3}$	S_6
$\bar{4}$	S_4
$\bar{6}$	S_3
Reflection	
m	σ
Inversion	
$\bar{1}$	i

the X-ray beam, for reasons we shall see later. There are many different kinds of instrument for recording X-rays scattered by crystals, and they produce a variety of appearances, but in each case a good quality crystal always gives a pattern of spots of varied intensity. In Fig. 1.13, different intensities are represented by different sizes of spot; on a photographic film, they would vary in their degree of blackness.

The pattern of spots has three properties of interest, which correspond to three properties of the crystal structure.

First, the pattern has a particular *geometry*. The spots lie in certain positions which are clearly not random. Each spot is generated at the detector by an individual scattered X-ray beam travelling in a definite direction from the crystal. This pattern geometry is related to the lattice and unit cell geometry of the crystal structure and so can tell us the repeat distances between molecules.

Second, the pattern has *symmetry*, not only in the regular arrangement of the spots but also in having equal intensities of spots which lie in symmetry-related positions relative to the centre of the pattern. The pattern symmetry is closely related to the symmetry of the unit cell of the crystal structure, i.e. to the crystal system and space group. The pattern in Fig. 1.13 has both vertical and horizontal reflection symmetry, and an inversion point at the centre.

Third, apart from this symmetry, there is no apparent relationship among the *intensities* of the individual spots, which vary widely; some are very intense, while others are too weak to be seen (their positions are deduced from the regular array). These intensities hold all the available information about the positions of the atoms in the unit cell of the crystal structure, because it is the relative atomic positions which, through the combination of their individual interactions with the X-rays, generate different amplitudes for different directions of scattering.

Thus, measurement of the geometry and symmetry of an X-ray scattering pattern provides information on the unit cell geometry and symmetry, while determination of the full molecular structure involves the measurement of all the many individual intensities, a considerably longer task.

An understanding of these relationships is valuable in grasping the fundamentals of X-ray diffraction in crystal structure determination. They can be summarized elegantly in mathematical terms, and the main equations will be presented shortly, but a pictorial approach may be more informative to many readers and will be given first. To illustrate the relationships between objects and their scattering patterns simply, we restrict the treatment to two-dimensional objects and scale up the process by a factor of about 10^4, so that atoms in planar molecules are represented by small holes punched in opaque card and monochromatic X-rays are replaced by monochromatic light with a wavelength comparable to the hole size and spacing. By doing this, it is possible to obtain from a modified type of microscope both an image (a picture) of each object and its light scattering pattern (also known as its optical transform) for comparison. Twelve pairs of objects and corresponding optical transforms are shown in Fig. 1.14.

Note first that every object produces a different pattern. These patterns of light and dark are generated by interference effects, i.e. by the combination of light waves coming from different parts of the object as the incident light passes through it.

Even for a single circular hole (think of this as a single atom in an X-ray beam) there are interference effects for the light waves scattered by the edges of the hole (objects **A** and **B**). In some directions these waves are in phase and the scattering pattern is bright; in other directions, scattered waves are out of phase and cancel each other, so little or no net intensity is seen. Thus the transmitted light does not give a sharp pattern matching the shape and size of the hole, but a diffuse pattern with a central circular bright region surrounded by rings of lower intensity. A larger hole (**B**) gives a smaller pattern. A rectangular hole (**C**) gives a diffuse pattern of rectangular symmetry, also with light and dark fringes. Note that the relative dimensions of the rectangle (tall and narrow) are reversed in the scattering pattern (short and wide). A series of experiments with different rectangles shows that there is an exact inverse relationship.

Several holes together represent a single molecule (**D** to **F**). These give more complicated scattering patterns, each with diffuse areas of varied intensity. There are now additional scattered wave interference effects, which depend on the relative positions of the holes. The rectangular object (**D**) gives a pattern with rectangular symmetry. The regular hexagon of holes (**F**) gives a pattern with the same six-fold symmetry. The object with only one vertical reflection line of symmetry (**E**), however, generates a pattern with extra symmetry: two mutually perpendicular lines of reflection, horizontal and vertical, intersecting in an inversion point. In general, each optical transform has the same symmetry as the corresponding object, with the addition of inversion symmetry if it is not already present (and this may imply further symmetry elements as in this case); the scattering pattern never has less symmetry than the object. In three dimensions, an equivalent rule applies, with the addition of an inversion centre to all scattering patterns.

Parts **G** to **I** show the effect of pure translation symmetry on the scattering of radiation by an object. Again, extra interference effects take place for the light rays scattered from individual holes. Because the holes are regularly spaced, these interference effects strongly reinforce each other, and the most obvious result is that the scattering patterns contain sharp maxima of intensity instead of broad diffuse regions. A single row of holes gives a pattern of narrow bright stripes running perpendicular to the row (**G** and **H**). A two-dimensional regular array of holes (**I**) imposes a restriction on the intensity maxima in both dimensions simultaneously, so that only sharp points of light are now seen (where two perpendicular sets of stripes cross), and these are also regularly spaced. The rows of spots in the optical transform always lie perpendicular to the rows of holes in the object, and there is an inverse relationship in the spacings: the array of holes in the object **I** are spaced wider horizontally than vertically, and the bright spots in the scattering pattern are spaced wider vertically than horizontally, with the inverse ratio.

The extra bright fringes and spots near the centre of these patterns are a result of the small number of holes in the objects. With more holes, these subsidiary maxima decrease in intensity and eventually disappear, and the main maxima become sharper. A real crystal may contain millions of unit cells in each direction, so the maxima for scattered X-rays are sharp.

This effect is well-known and exploited in many branches of science, and is called *diffraction*. Here we see that a regular lattice arrangement of objects scattering radiation produces severe diffraction restrictions, so that the scattered radiation has significant intensity only in certain well-defined directions and not in a diffuse pattern such as occurs with a single object.

Finally, we place more complicated objects (hexagons, representing molecules, instead of single holes) on lattices. The three lattices **J** to **L** have different geometries (different unit cells) but each has as its basic repeat unit the same single hexagon in the same orientation (which is the same as **F**, but turned through 90°). Each produces a diffraction pattern consisting of a regularly spaced array of more or less bright points. The positions of these points are dictated by the diffraction conditions, generated by the parent lattice: in each case, the rows of points run perpendicular to the rows of hexagons on the object lattice, and the spacing of the points in each direction is inversely proportional to the spacing of the hexagons. The intensities of the individual spots are produced by the form of the single object (the hexagon): comparison of **J**, **K** and **L** with **F** (especially with half-closed eyes to blur the lattice diffraction effects!) shows that the underlying pattern of light and dark is the same for all of them, and this is the optical transform of the single hexagon.

The net effect is like looking at the optical transform of the hexagon through a sieve. The transform itself (the variation of intensity) is determined by the form of the single object, while the mesh of the sieve, which dictates the points at which the transform intensities can actually be seen, is determined by the lattice geometry.

Extending this to three dimensions, translating it to the case of a single crystal in a monochromatic X-ray beam, and introducing some formal terminology, these relationships illustrated pictorially are summarized as follows. An object scatters radiation of wavelength comparable to its own size; the mathematical relationship between the object and the scattering pattern is *Fourier transformation*, such that the scattering pattern is the *Fourier transform* of the object, and the image of the object (provided by an optical microscope or by X-ray crystallography) is in turn the Fourier transform of the scattering pattern. If identical objects are arranged on a lattice, diffraction effects of the lattice are also imposed, so that the diffraction pattern can have non-zero intensity only where the direction of scattering satisfies the equations for diffraction geometry. The overall effect is a combination of the two effects – scattering by the object further restricted by diffraction by the lattice – so the

Fourier transformation is a well-known and well established mathematical operation used in a wide variety of science and technology applications, including spectroscopy and image processing.

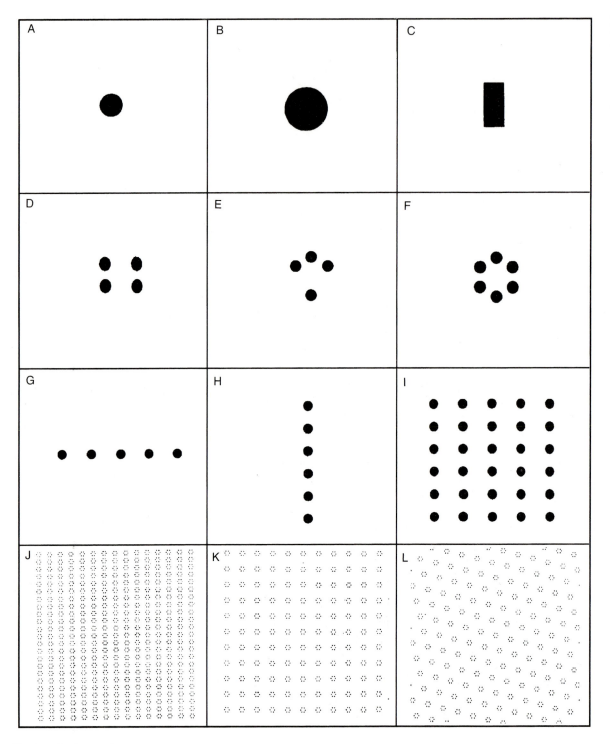

Fig. 1.14   A selection of objects and their optical scattering patterns. Each of the 12 objects on the left page has its scattering pattern in the corresponding position on the right page. The bottom row of objects is on a different scale; in the objects themselves, each hexagon has the same size as the individual hexagon in the second row. In the description in the text, the objects and their patterns are labelled from (**A**) to (**C**) in the top row, continuing to (**L**) at the end of the bottom row. These illustrations are taken from a very comprehensive compilation in *Atlas of Optical Transforms* by G. Harburn, C. A. Taylor and T. R. Welberry (G. Bell and Sons Ltd, London, 1975), with the permission of the authors.

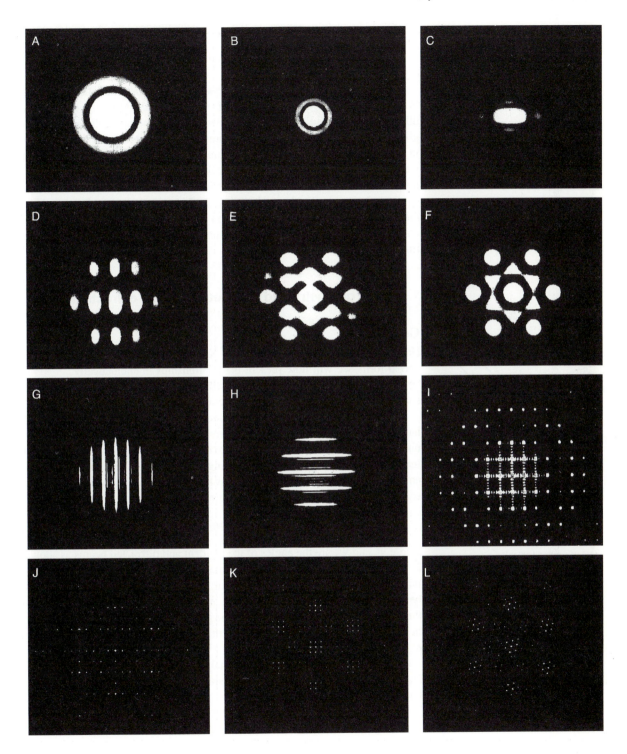

Fig 1.14 *continued*

observed diffraction pattern is the Fourier transform of the single object sampled at certain geometrically determined points.

The previous paragraph is the essential basis of the technique of X-ray crystallography.

In principle, then, the process of crystal structure determination is simple. We record the diffraction pattern from a crystal. Measurement of the diffraction pattern geometry and symmetry tells us the unit cell geometry and gives some information about the symmetry of arrangement of the molecules in the unit cell. Then from the individual intensities of the diffraction pattern we work out the positions of the atoms in the unit cell, by pretending to be a microscope lens system, adding together the individual waves with their correct relative amplitudes and phases. And here we see the *phase problem*, the fact that the measured diffraction pattern provides directly only the amplitudes and not the required phases, without which the Fourier transformation cannot be made.

1.5   The geometry and symmetry of X-ray diffraction

Geometry

Having seen the fundamental basis of X-ray diffraction in both pictorial and verbal form, we will now present the relationships mathematically.

For the geometry, consider first diffraction by a single row of regularly spaced points (one-dimensional diffraction; Fig. 1.15).

In any particular direction, the radiation scattered by the row of points will have zero intensity by destructive interference of the individual scattered rays unless they are all in phase. Since, except in the straight-through direction, individual rays have different path lengths, these path differences must be equal to whole numbers of wavelengths to keep the rays in phase. So, for rays scattered by two adjacent points in the row

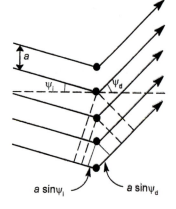

Fig. 1.15   Diffraction by a single row of regularly spaced objects.

$$\text{path difference} = a \sin \Psi_i + a \sin \Psi_d = h\lambda \qquad (1.1)$$

where Ψ_i and Ψ_d are the angles of the incident and diffracted beams as shown, λ is the wavelength, a is the one-dimensional lattice spacing, and h is an integer (positive, zero, or negative). For a given value of Ψ_i (a fixed incident beam), each value of h corresponds to an observed diffraction maximum and the equation can be used to calculate the permitted values of Ψ_d, the directions in which intensity is observed. The result, as we have seen in Fig. 1.14, is a set of bright fringes.

For diffraction by a three-dimensional lattice there are three such equations and all have to be satisfied simultaneously. The first equation contains the lattice a spacing, angles relative to this a axis of the unit cell, and an integer h. The other two equations, correspondingly, contain the unit cell axes b and c, and integers k and l respectively.

Thus each allowed diffracted beam (each spot seen in an X-ray diffraction pattern) can be labelled by three integers, or indices, *hkl*, which uniquely specify it if the unit cell geometry is known.

These three equations for diffraction geometry, the *Laue conditions*, are cumbersome to use in this form. An alternative but equivalent description was

The letters *h*, *k*, and *l* are used conventionally by all crystallographers although, unlike other conventional triplets of letters used in the subject (such as *a*, *b*, *c*; *x*, *y*, *z*), they are not consecutive in the alphabet.

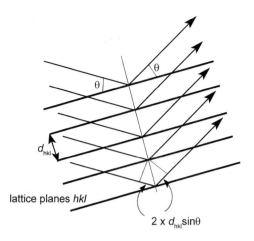

Fig. 1.16 The Bragg construction for diffraction by a three-dimensional crystal structure; one set of parallel lattice planes is seen edge-on.

derived by W. L. Bragg soon after the experimental demonstration that X-rays could be diffracted by crystals, and is expressed in the single *Bragg equation*, which is universally used as the basis for X-ray diffraction geometry (Fig. 1.16). Bragg showed that every diffracted beam that can be produced by an appropriate orientation of a crystal in an X-ray beam can be regarded geometrically as if it were a reflection from sets of parallel planes passing through lattice points, analogous to the reflection of light by a mirror, in that the angles of incidence and reflection must be equal and that the incoming and outgoing beams and the normal to the reflecting planes must themselves all lie in one plane. The reflection by adjacent planes in the set gives interference effects equivalent to those of the Laue equations; to define a plane we need three integers to specify its orientation with respect to the three unit cell edges, and these are the indices *hkl*; the spacing between successive planes is determined by the lattice geometry, so is a function of the unit cell parameters.

At the age of 24 W. L. Bragg, together with his father W. H. Bragg, was awarded the Nobel Prize for Physics for this work and its application in the first crystal structure determinations. Max von Laue received the Prize previously for his part in the discovery of the diffraction of X-rays by crystals.

For rays reflected by two adjacent planes

$$\text{path difference} = 2d_{hkl} \sin \theta = n\lambda \qquad (1.2)$$

In practice, the value of n can always be set to 1 by considering planes with smaller spacing ($n = 2$ for planes *hkl* is equivalent to $n = 1$ for planes $2h$, $2k$, $2l$ which have exactly half the spacing) and it is in the form

$$\lambda = 2d_{hkl} \sin \theta \qquad (1.3)$$

that the Bragg equation is always used. It allows each observed diffracted beam (commonly known as a 'reflection') to be uniquely labelled with its three indices and for its net scattering angle (2θ from the direct beam direction) to be calculated from the unit cell geometry, of which each d_{hkl} spacing is a function.

Rearrangement of the Bragg equation gives

$$\sin \theta = \left(\frac{\lambda}{2}\right) \times \left(\frac{1}{d_{hkl}}\right) \qquad (1.4)$$

The distance of each spot from the centre of an X-ray diffraction pattern such as Fig. 1.13 is proportional to $\sin \theta$ and hence to $1/d_{hkl}$ for some set of lattice planes. This demonstrates mathematically the reciprocal (inverse) nature of the geometrical relationship between a crystal lattice and its diffraction pattern, already seen pictorially.

The Bragg equation is the basis of all methods for obtaining unit cell (lattice) geometry from the measured geometry of the diffraction pattern. The exact application depends on the experimental setup used to obtain the diffraction pattern.

Symmetry

Simple rotation and reflection symmetry is seen directly in the diffraction pattern, always with the addition of an inversion centre if it is not already there. Glide planes and screw axes cause particular subsets of reflections to have exactly zero intensity, according to well-established rules for these 'systematic absences'. The effect can be seen in the central rows, both horizontal and vertical, in Fig. 1.13, where alternate reflections have zero intensity.

The unit cell dimensions provide some information about the distances between molecules, which are regularly spaced in the crystal structure. In most structures, however, each unit cell contains not one but several molecules which are related to each other by the space group symmetry. This symmetry is revealed in various aspects of the appearance of the diffraction pattern, from which it is usually possible to choose the correct space group from the complete list of 230 or, at least, to narrow down the choice to a few possibilities; ambiguities arise, for example, because a diffraction pattern may have more symmetry than the structure itself, but it cannot have less, and in such cases the correct answer is known only when the structure is successfully solved and refined.

For a compound of known chemical formula, the number of molecules per unit cell can be calculated (if the density of the crystals is measured or estimated, as illustrated below). This number can be compared with the number of 'asymmetric units' required by the symmetry elements present in the space group (see section 1.3). If the two are equal, then there is one molecule per asymmetric unit and this tells us nothing about the molecular shape. If, however, the asymmetric unit is only a fraction of a molecule, then the molecule itself must have one or more symmetry elements of the space group, and this provides some information on the molecular shape, even before the full structure determination is carried out.

This is best illustrated with examples. The first is rather unusual, with a very high symmetry. Crystals of $[Cr(NH_3)_6][HgCl_5]$, obtained from aqueous solution, belong to the cubic crystal system, with $a = 22.653$ Å. The unit cell volume is $V = abc = a^3 = 11\,625$ Å^3. Crystal densities can be measured by various methods; for this compound, a value of $D = 2.44$ g cm^{-3} is obtained.

Since the density and unit cell volume are known experimentally, the mass of the contents of one unit cell can be calculated.

Be careful of the units in these calculations; they need to be consistent, and initial data usually have to be multiplied or divided by some powers of 10 to achieve this. For simplicity, uncertainties in the experimental measurements are ignored here; this topic is discussed later.

$$\text{mass} = \text{density} \times \text{volume}$$

$$\begin{aligned} \text{so, unit cell mass} &= 2.44 \text{ g cm}^{-3} \times 11\,625 \text{ Å}^3 \\ &= 2.44 \text{ g cm}^{-3} \times 11\,625 \times (10^{-8})^3 \text{ cm}^3 \\ &= 2.837 \times 10^{-20} \text{ g per unit cell} \end{aligned} \tag{1.5}$$

To convert between grams for one unit cell (or for one molecule) and atomic mass units (officially called daltons in SI; these are masses in grams for one mole), the scale factor is Avogadro's number.

$$\begin{aligned} \text{unit cell mass} &= 2.837 \times 10^{-20} \times 6.023 \times 10^{23} \\ &= 17\,090 \text{ daltons} \end{aligned} \tag{1.6}$$

The mass of one formula unit is just the sum of all the atomic masses, in this case 532.0 daltons. From the known (or assumed!) formula mass and the experimentally determined total unit cell mass, the ratio gives the number of formula units ('molecules') per unit cell, conventionally given the symbol Z.

The term 'molecular mass' is generally used, though this particular compound is ionic rather than molecular; 'mole mass' is more appropriate.

$$Z = \text{unit cell mass/formula mass}$$
$$= 17\,090/532.0 \qquad (1.7)$$
$$= 32.1$$

This must be a whole number and appropriate to the symmetry of the space group, in this case $Fd\bar{3}c$ (number 228 of the 230). There must, therefore, be 32 complex cations and 32 complex anions in each unit cell. Reference to the space group tables shows that the asymmetric unit for space group $Fd\bar{3}c$ is 1/192 of the unit cell, so a molecule or ion with no symmetry of its own would have $Z = 192$. The cation and anion here have considerable symmetry themselves. According to the tables, for $Z = 32$, the allowed point group symmetries of the ions are S_6 and D_3. Of these, the cation must have S_6 symmetry, which is consistent with essentially octahedral coordination of the Cr by six NH_3 ligands, and the anion has D_3 symmetry, which means it is a regular trigonal bipyramid with two axial and three equatorial ligands attached to Hg.

This rather extreme case shows that sometimes a great deal can be deduced about the molecular shape, even before the full structure determination is carried out.

The subject of space groups is beyond the scope of this book. The space group symbol shows the type of unit cell centring (F means all-face-centred) and the combination of symmetry elements given as lower-case letters (mirror planes and glide planes) and numbers (rotation axes and screw axes). Some of the simpler space group symbols are explained more fully in later examples.

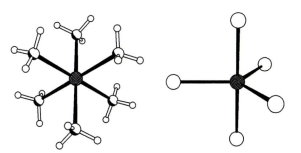

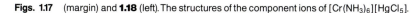

Figs. 1.17 (margin) and **1.18** (left). The structures of the component ions of $[Cr(NH_3)_6][HgCl_5]$.

As a second example, the complex $[(C_4H_9)_4N]_2[WO_2Cl_4]$ (formula mass 842.6) crystallizes from solution in 1,2-dichloroethane ($ClCH_2CH_2Cl$) to give monoclinic crystals, space group $P2_1/n$, with $a = 15.821$, $b = 17.117$, $c = 18.475$ Å, $\beta = 101.14°$. If all cell angles were 90°, as in the cubic example above, the cell volume would be $V = abc$; with just one non-90° angle, the volume is $V = abc\cdot\sin(\text{angle})$, here $abc\cdot\sin\beta$, which amounts to 4908.9 Å^3. The density $D = 1.40$ g cm^{-3}. Following the same procedure as for the first example,

$$\text{unit cell mass} = \text{density} \times \text{cell volume}$$
$$= 1.40 \times 4908.9 \times 10^{-24} \times 6.023 \times 10^{23}$$
$$= 4139 \text{ daltons} \qquad (1.8)$$
$$Z = \text{unit cell mass/formula mass}$$
$$= 4139/842.6$$
$$= 4.91$$

The space group $P2_1/n$ (an alternative setting for the conventional $P2_1/c$, with a different choice of unit cell axes) is, in fact, the most common of all 230, accounting for roughly one-third of all known molecular crystal structures; its arrangement of symmetry elements provides a particularly effective packing of molecules of general shape. Fortunately, it is one of the space groups which gives a unique set of systematic absences in the diffraction pattern. The screw axis, parallel to the b axis, causes reflections $0k0$ to be absent when k is odd; the glide plane, perpendicular to the b axis and with its glide direction along the ac face diagonal, causes reflections $h0l$ to be absent when $h + l$ is odd. The presence of the screw axis and the perpendicular n-glide plane are indicated in the space group symbol $P2_1/n$, the capital letter P indicating a primitive (not centred) unit cell.

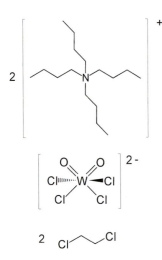

Fig. 1.19   The component ions and solvent molecules in the crystal structure of $[(C_4H_9)_4N]_2[WO_2Cl_4]\cdot 2ClCH_2CH_2Cl_2$.

The equation for a triclinic unit cell volume, involving all six cell parameters, is much more complicated than those given for the first two examples.

This presents a problem, since Z is not a whole number. The closest integer is 5, but $Z = 5$ is extremely unlikely for the monoclinic system, in which symmetry elements relate molecules in pairs, and in any case the experimental errors are not so large.

If the experimental measurements (unit cell geometry and density) are correct, the answer to the problem must lie in the chemical formula, which has been assumed, not proved. To find the true formula mass instead of the assumed one, we must choose an appropriate integral value for Z, in this case 4, and work backwards, from what we know to what we do not.

$$\text{formula mass} = \text{unit cell mass}/Z$$
$$= 4139/4 \tag{1.9}$$
$$= 1035 \text{ daltons}$$

This is 192.4 greater than the previously assumed formula mass and the difference is, within reasonable experimental error, equal to the molecular mass for $2ClCH_2CH_2Cl$ (197.9). The true complete formula of the compound is, therefore, probably $[(C_4H_9)_4N]_2[WO_2Cl_4]\cdot 2ClCH_2CH_2Cl$; for every two cations and one anion in the crystal structure, there are also two molecules of 1,2-dichloroethane solvent, incorporated during the crystallization. This gives a calculated density of 1.408 g cm^{-3}, in good agreement with the measured value.

Such solvent of crystallization is by no means uncommon, and is certainly not restricted to the familiar case of water of crystallization in many salts obtained from aqueous solution (hydrates).

The expected value of Z for space group $P2_1/n$ is 4, and here we have eight cations, four anions and eight solvent molecules in each unit cell, so there are no requirements for any of these constituents to have imposed symmetry. At this stage we can deduce nothing about the shapes of the ions from symmetry arguments; this is more often the case than not.

As a final example, which we shall consider further in Chapter 3, a compound $[(18\text{-crown-}6)K][In(SCN)_4(py)_2]$ (where py is pyridine) can be obtained from solution in pyridine (which serves not only as a solvent, but also as a ligand to the indium atom). It crystallizes in the triclinic system with $a = 8.941$, $b = 9.682$, $c = 13.113$ Å, $\alpha = 87.25$, $\beta = 72.33$, $\gamma = 89.05°$, giving $V = 1080.3$ Å^3. The compound is air-sensitive and there are other experimental difficulties in measuring the density. In fact, crystal densities are often not measured at all. Experience shows that, for a wide range of organic compounds and metal complexes with organic ligands, the average volume required for a molecule in a crystal structure is about 18 Å^3 per non-hydrogen atom (hydrogen atoms are not counted). The proposed formula for this compound has 44 non-hydrogen atoms, so a volume of about 792 Å^3 is expected for each cation–anion pair. This is rather less than the measured unit cell volume, but more than half of it, so the unit cell cannot contain more than one formula unit, and the proposed formula is incomplete (or incorrect): the cell volume is sufficient for one cation, one anion, and two or three molecules of pyridine. There are two possible triclinic space groups: $P1$, which has no symmetry other than pure translation, and $P\bar{1}$, which has inversion symmetry. Of these, the centrosymmetric space group $P\bar{1}$ is far more common, except for structures of chiral molecules. In this space group, the asymmetric unit is half the unit cell, the other half being related to it by inversion symmetry, so the expected value of Z is 2. Since this structure has $Z = 1$ (so there is only one cation and one anion in each unit cell), both the cation and the anion must themselves have symmetry, and must lie on inversion centres. Therefore, the $[(18\text{-crown-}6)K]^+$ cations and the $[In(SCN)_4(py)_2]^-$ anions are both centrosymmetric. In the case of the anion, this means that identical ligands must occur in pairs *trans* to each other, with the indium atom sitting on the inversion centre. Hence we already know (assuming the proposed chemical formula is correct apart from additional solvent molecules, and that the much more common triclinic space group applies) that the pyridine ligands are *trans* to each other, not *cis*, one of the questions to be answered by carrying out the structure determination.

There are, of course, numerous instances in which the proposed chemical formula and the experimentally measured unit cell volume are just not compatible, with no possible integer value of Z, even when possible solvent of crystallization is included. In such cases, these preliminary measurements and calculations show that the material being studied is simply not what was thought. Sometimes it is starting material or a decomposition product, but sometimes it is a totally unexpected and unknown material of considerable interest. Without any other non-crystallographic information, only a full structure determination based on the measured intensities will show the answer, unless the unit cell can be recognized as that of an already known crystal structure (perhaps with the help of a computer database, as described in Chapter 4).

Further examples of these calculations, in the form of problems for solution by the reader, can be found at the end of this chapter (section 1.8).

1.6   The intensities of diffracted X-rays

Background and notation

The intensities of the diffraction pattern and the arrangement of atoms in the unit cell of the crystal structure are related to each other by Fourier transformation: the diffraction pattern is the Fourier transform of the electron density, and the electron density is itself the Fourier transform of the diffraction pattern.

The mathematical equations for crystallographic Fourier transformations have a fearsome appearance, but they can be understood in terms of the physical processes which they represent, with the help of the optical analogues presented earlier. Much of the difficulty presented by the Fourier transform equations comes from their use of complex number notation. This can be regarded as just a convenient way of manipulating two numbers with only one symbol. The two numerical values associated with each reflection in a crystal diffraction pattern are the *amplitude* $|F|$ and the phase ϕ of the diffracted wave. Figure 1.21 shows two such waves; the amplitude $|F|$ is represented by the height of the wave, and the phase ϕ by the horizontal shift relative to some chosen origin.

Another, more compact, way of representing the same waves is shown in Fig. 1.22. Each wave is represented by an arrow with its tail at the centre of the diagram (the origin); the length of the arrow is proportional to the wave

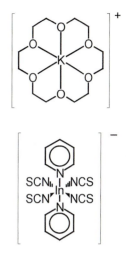

Fig. 1.20   The cation and anion of $[(18\text{-crown-}6)K][In(SCN)_4(py)_2]$.

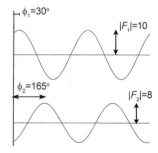

Fig. 1.21   Amplitudes and phases of two waves.

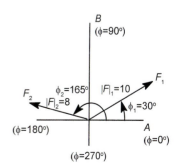

Fig. 1.22   The same two waves as in Fig. 1.21, represented as vectors.

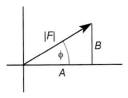

Fig. 1.23 The relationship between vector and coordinate representations.

amplitude $|F|$, and the direction shows the phase ϕ, with a zero phase angle on the horizontal axis to the right and other angles (0–360° or 0–2π radians) measured anti-clockwise from there. This is a vector representation: a vector **F** has both magnitude $|F|$ and direction ϕ, like the arrows in the diagram.

Instead of the two values of length and direction from the origin, each of the arrowhead positions could be specified by two coordinates on the horizontal (A) and vertical (B) axes. The mathematical relationship between the vector and coordinate representations is in terms of the Pythagoras theorem and simple trigonometry (Fig. 1.23).

$$|F|^2 = A^2 + B^2; \quad \tan\phi = B/A$$
$$A = |F|\cos\phi; \quad B = |F|\sin\phi \tag{1.10}$$

Forming an image of electron density from a diffraction pattern is the equivalent of the operation of a microscope lens system and involves adding together waves with their correct relative amplitudes and phases. This is shown in wave terms for just two waves in Fig. 1.24 and in vector terms in

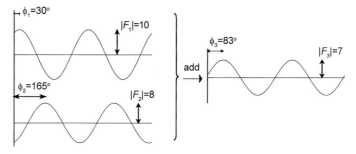

Fig. 1.24 Addition of two waves to give a resultant wave.

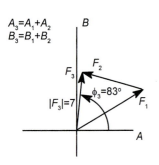

Fig. 1.25 The same wave addition as in Fig. 1.24, as a vector representation.

Fig. 1.25. The A component of the resultant combined vector is simply the sum of the A components of the individual vectors, and similarly for the B components. Then the final amplitude $|F|$ and phase ϕ can be calculated from the final A and B by equations 1.10. This is true for the combination of any number of waves.

$$\text{combined } A = A_1 + A_2 + \cdots + A_n = \sum_{i=1}^{n} A_i$$
$$\tag{1.11}$$
$$\text{combined } B = B_1 + B_2 + \cdots + B_n = \sum_{i=1}^{n} B_i$$

Clearly, the A and B components must be summed separately and not mixed up together during the process until the sums are complete. The A components are terms involving cosines of phase angles, and the B components are analogous terms involving sines of phase angles (equation 1.10).

In practice, computer programs to calculate crystallographic Fourier transforms do treat the A and B components of the individual reflections separately in this way. For convenience in showing the mathematical relationships, however, avoiding the need for two versions of every equation, the two components can be represented by a single symbol using complex number notation. A complex number has two parts, which are kept separate by multiplying one of them by the symbol i. A full treatment of complex number theory is beyond the scope or requirement of our subject. Here we need only a few of its most important features. The 'non-i-terms' and the 'i-terms' are equivalent to two orthogonal coordinates, the components of two-dimensional vectors (the horizontal and vertical axes in Figs 1.22 and 1.25). Multiplication by i is equivalent to rotating a vector by 90° anticlockwise, so multiplying by i^2 is a 180° rotation which turns a vector $\mathbf{F}$ into its opposite vector $-\mathbf{F}$. So we can write one symbol F for a wave, where

$$F = A + iB \qquad (1.12)$$

From the previous relationship between the vector and coordinate representations, and using a property of complex numbers whereby $e^{i\phi} = \cos\phi + i\sin\phi$, the above equation becomes

$$\begin{aligned} F &= |F|\cos\phi + i|F|\sin\phi \\ &= |F|(\cos\phi + i\sin\phi) \qquad (1.13) \\ \text{so } F &= |F| \cdot e^{i\phi} \end{aligned}$$

This means, logically, that i is the square root of -1, a difficult concept which leads to the unhelpful and misleading description (as far as our subject is concerned) of the two components as 'real' and 'imaginary': they are, in fact, both equally real!

and we have the amplitude $|F|$ and phase ϕ represented by the one symbol F, a complex number. Remember that each reflection, or diffracted wave, is labelled by its three indices hkl, so for each reflection

To avoid cramped superscripts, $e^{i\phi}$ can also be written as $\exp(i\phi)$. We adopt this notation from here on.

$$F(hkl) = |F(hkl)| \cdot \exp[i\phi(hkl)] \qquad (1.14)$$

$F(hkl)$ is called the *structure factor* of the reflection with indices h, k, and l.

The forward Fourier transform (the diffraction experiment)

The diffraction pattern is the *Fourier transform* (FT) of the electron density. In mathematics:

$$F(hkl) = \int_{\text{cell}} \rho(xyz) \cdot \exp[2\pi i(hx + ky + lz)]\mathrm{d}V \qquad (1.15)$$

The structure factor (amplitude and phase) for reflection hkl is given by taking the value of the electron density at each point in the unit cell, $\rho(xyz)$, multiplying it by the complex number $\exp[2\pi i(hx + ky + lz)]$, and adding up (integrating over the whole cell volume, $\int_{\text{cell}} \cdots \mathrm{d}V$) these values. Positions in the unit cell are measured from one corner (the origin) and the coordinates x, y, z are in fractions of the corresponding cell edges a, b, c: for example, the very centre of the unit cell has coordinates $x = \frac{1}{2}$, $y = \frac{1}{2}$, $z = \frac{1}{2}$. This calculation can be carried out mathematically to mimic the observed experimental diffraction of X-rays by a crystal. It needs to be done for each reflection and it produces a set of calculated structure factors, each with an amplitude $|F(hkl)|$ and a phase $\phi(hkl)$. In the experiment itself, of course, only the amplitudes are obtained.

This equation shows how each bit of the structure contributes to every reflection. Since all the unit cells are identical, the total diffraction pattern of the crystal is just the Fourier transform of the contents of one unit cell multiplied by the number of unit cells in the crystal, so there should be just a simple scale factor between the observed and calculated sets of amplitudes.

The equation in this form is not convenient for calculation, because it contains integration and a continuous function $\rho(xyz)$. Summation of a finite number of terms is easier. This can be achieved by expressing the electron density instead in terms of individual atoms.

One atom scatters X-rays rather like a single circular hole scatters light passing through it (Fig. 1.14), except that the scattering is by electrons

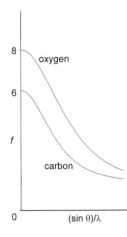

Fig. 1.26   X-ray atomic scattering factors for carbon and oxygen.

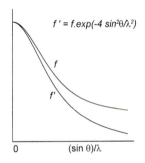

Fig. 1.27   The effect of atomic vibration on X-ray scattering factors; in this example, $8\pi^2 U = 4$ Å^2.

Note that there is a diffracted wave $F(000)$, for which $\theta = 0$ and so $|F(000)|$ is the sum of all the zero-angle atomic scattering factors; this is just the total number of electrons in one unit cell. The intensity of this wave cannot be measured experimentally, because it coincides with the majority of the incident X-ray beam, which passes undeflected through the crystal. All other $|F(hkl)|$ values are smaller than $|F(000)|$, which represents all the atoms scattering together cooperatively.

throughout the atom and not just on its edges; this means no outer rings of brightness are formed. In the forward direction ($2\theta = 0°$) all the electrons scatter X-rays exactly in phase, but at all other angles there are partial destructive interference effects, so the intensity falls off as θ increases. The variation of intensity with angle (usually shown as a function of $(\sin\theta)/\lambda$ so that it is the same for X-rays of different wavelengths) is called the atomic scattering factor $f(\theta)$ and has the general form shown in Fig. 1.26. It is measured in units of electrons; $f(0)$, the scattering factor for zero deflection, is equal to the atomic number. These functions are known for atoms and ions of all elements and are obtained from quantum mechanical calculations; they are available in standard reference tables and incorporated into many crystallographic computer programs.

Atoms in crystalline solids, however, are not stationary; they vibrate, to an extent which depends on the temperature, and this effectively spreads out the atomic electron density and increases the interference effects. The atomic scattering factor falls off more rapidly with increasing angle, and is not the same for all atoms of the same element, because they generally have different amounts of vibration unless they are symmetry-equivalent. For an atom which vibrates equally in all directions (isotropic vibration), the effect is to multiply the atomic scattering factor by a term containing an *isotropic displacement parameter* U (see Fig. 1.27), which represents a mean-square amplitude of vibration for the atom, a measure of how much it is vibrating.

$$f'(\theta) = f(\theta) \cdot \exp\left(-\frac{8\pi^2 U \sin^2 \theta}{\lambda^2}\right) \qquad (1.16)$$

Note that U has units Å^2 and the extra term has a value < 1. The larger the value of U, the more the curve falls off at higher Bragg angles.

With discrete atoms instead of a continuous electron density function, the forward Fourier transform takes the form

$$F(hkl) = \sum_j f_j(\theta) \cdot \exp(-8\pi^2 U_j \sin^2 \theta/\lambda^2) \cdot \exp[2\pi i(hx_j + ky_j + lz_j)]$$
$$\qquad (1.17)$$

1   **2**   **3**   **4**   **5**

The summation is made over all the atoms in the unit cell, each of which has its appropriate atomic scattering factor f_j (a function of the Bragg angle θ), a displacement parameter U_j, and coordinates (x_j, y_j, z_j) relative to the unit cell origin. This summation must be carried out for every diffracted wave $F(hkl)$.

Although equation 1.17 looks complicated, it can be readily understood in terms of the physical process it represents. Every atom scatters X-rays falling on it (terms **3** and **4** in the equation). In any particular direction (hkl), these separate scattered waves from each atom have different relative phases which depend on the relative positions of the atoms (term **5**), and the total diffracted wave in that direction (term **1**) is just the resultant sum (term **2**) of the X-rays scattered by the individual atoms. The equation just represents the combination or addition of many waves to give one resultant wave in each direction; the graphical equivalent for two waves was given in Fig. 1.24.

The reverse Fourier transform (the recombination calculation)

The electron density is the *reverse Fourier transform* (**FT**$^{-1}$) of the diffraction pattern. Because the diffraction pattern of a crystal consists of discrete reflections rather than a diffuse pattern, this Fourier transform is a summation, not an integral.

$$\rho(xyz) = \frac{1}{V}\sum_{h,k,l} F(hkl)\cdot \exp[-2\pi i(hx+ky+lz)]$$

$$\text{or } \rho(xyz) = \frac{1}{V}\sum_{h,k,l}|F(hkl)|\cdot\exp[i\phi(hkl)]\cdot\exp[-2\pi i(hx+ky+lz)] \quad (1.18)$$

$$\underset{1}{}\qquad\underset{2}{}\;\underset{3}{}\qquad\underset{4}{}\qquad\quad\underset{5}{}$$

Remember that $F(hkl)$ is a complex number, containing both amplitude and phase information, as is shown explicitly in the second version. The term $1/V$ is necessary in order to give the correct units (structure factors, like atomic scattering factors, have units of electrons, but electron density is electrons per Å^3.

The summation is performed over all values of h, k, and l, i.e. all the reflections in the diffraction pattern contribute to it. In practice, reflections are measured only to a certain maximum Bragg angle, but this is usually unimportant because the higher angle reflections are weaker and so contribute relatively little to the sums. The summation must be carried out for many different coordinates x, y, z in order to show the variation of electron density in the unit cell and hence locate the atoms where the electron density is concentrated in peaks.

As for the forward Fourier transform, this equation is readily understood in terms of the (unachievable!) physical process it represents. The image of the electron density (**1**), which originally generated the diffraction pattern, is obtained by adding together (**2**) all the diffracted beams, with their correct amplitudes (**3**) and phases (**4,5**); the correct relative phases here include the intrinsic phases of the waves themselves, relative to the original incident beam (**4**), and an extra phase shift appropriate to each geometrical position in the image relative to the unit cell origin (**5**).

The relative phase shifts (**5**) can be calculated as required, but the intrinsic phases $\phi(hkl)$ of the different reflections are unknown from the diffraction experiment. This means that it is not possible simply to calculate the reverse Fourier transform once the diffraction pattern has been measured. Here, once again, in the mathematical basis of the method we see the nature of the 'phase problem'.

1.7 Sources of X-rays

So far, uses of X-rays have been discussed, but nothing about how they are produced. Details are not important for our purposes, and only a brief outline is given here for completeness.

In most laboratories the standard source of X-rays is an '*X-ray tube*' (Fig. 1.28). This is an evacuated enclosure of glass and metal construction which produces electrons by passing an electrical current through a wire filament, accelerates them to a high velocity by an electrical potential of typically 40 000–60 000 volts across a few millimetres, then stops them dead with a water-cooled metal block. Most of the electron kinetic energy is converted to heat and wasted, but a small proportion generates X-rays by interaction with the target metal atoms. Among other effects occurring, if an electron in a core atomic orbital is ejected (ionized), an electron from a higher orbital can take its

The forward and reverse Fourier transforms differ in that one has a negative sign inside the exponential term.

The unmeasured term $F(000)$ must also be included; it is equal to the total number of electrons in one unit cell.

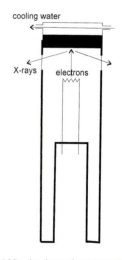

Fig. 1.28 A schematic representation of a laboratory X-ray tube.

Here, *h* is Planck's constant, not to be confused with *hkl* indices of X-ray reflections!

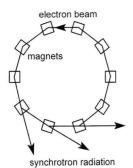

Fig. 1.29 The spectrum of X-rays emitted by an X-ray tube.

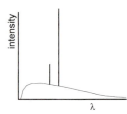

Fig. 1.30 The production of synchrotron radiation from relativistic electrons in a special type of particle accelerator.

place and the drop in energy produces emission of radiation of a definite frequency and wavelength ($\Delta E = h\nu = hc/\lambda$). Several such transitions are possible, so the output of radiation from the target consists of a series of intense sharp maxima, superimposed on a broad-spectrum background of radiation from non-quantum processes (Fig. 1.29). One particular peak, usually the most intense, can be selected and the rest of the output suppressed by exploiting the Bragg equation: the beam of radiation falls on a single crystal of known structure (often graphite) suitably oriented so that the desired wavelength satisfies the equation; only this wavelength is diffracted at the appropriate angle, the others pass straight through the monochromator crystal.

The most commonly used X-ray tube target materials are copper and molybdenum, which give characteristic X-rays of wavelengths 1.54184 and 0.71073 Å respectively.

Various developments of the basic X-ray tube produce higher intensities. The main limitation is the amount of heat produced. For a more powerful electron beam, the target must be constantly moved in its own plane to spread the heat load, producing 'rotating anode' X-ray tubes, which can provide about one order of magnitude more intensity.

Much more intense X-rays, as well as other parts of the electromagnetic spectrum, are produced in a *synchrotron storage ring* (Fig. 1.30), in which electrons moving at almost the speed of light are constrained by magnetic fields to move in a circle many metres in diameter. The radiation, emitted tangentially from the ring, has a continuous spectrum, ranging from infra-red to X-rays, from which a single wavelength of any value can be selected by a monochromator (the X-rays are 'tuneable') and is many orders of magnitude more intense than the output of an X-ray tube. Such sources, of course, are vastly more expensive and are national or international facilities with a wide variety of other scientific applications in addition to X-ray diffraction.

1.8   Problems

The following calculations of unit cell contents provide practice problems similar to the examples in section 1.5. Worked answers are given in the Appendix.

Problem 1.1

The complex $[(C_{18}H_{18}N_4S)HgBr_2]$ (relative molecular mass 682.8) crystallizes from solution in acetonitrile (CH_3CN) to give triclinic crystals with a unit cell volume of 1113.5 Å^3 and with a density of 2.16 g cm^{-3}. Calculate the number of molecules of complex per unit cell and the number of molecules of solvent per unit cell.

Problem 1.2

A compound of supposed formula $K^+[In(NCS)_4(bipy)]^-$, where bipy is the chelating ligand 2,2'-bipyridyl, is obtained from solution in thf (C_4H_8O) as monoclinic crystals with $a = 14.985$ Å, $b = 17.375$ Å, $c = 16.437$ Å, $\beta = 92.23$ Å. The density of the crystals is 1.40 g cm^{-3}, and the relative molecular mass for the above formula is 542.4. Calculate the unit cell volume and deduce the number of cations and anions per unit cell (expected values are 2, 4 or 8) and the number of molecules of thf per cation (which are likely to be coordinated to it).

2 The method step by step

2.1 A schematic flowchart

Having examined the physical basis of X-ray crystallography and its expression in mathematical notation, we consider now how the method works in practice. In this chapter the various successive steps are described in general terms with appropriate examples; in the next chapter several case studies are presented to illustrate the complete procedure.

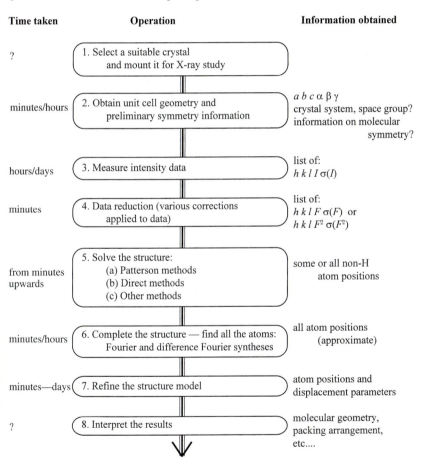

Time taken	Operation	Information obtained

? — 1. Select a suitable crystal and mount it for X-ray study

minutes/hours — 2. Obtain unit cell geometry and preliminary symmetry information — $a\ b\ c\ \alpha\ \beta\ \gamma$ crystal system, space group? information on molecular symmetry?

hours/days — 3. Measure intensity data — list of: $h\ k\ l\ I\ \sigma(I)$

minutes — 4. Data reduction (various corrections applied to data) — list of: $h\ k\ l\ F\ \sigma(F)$ or $h\ k\ l\ F^2\ \sigma(F^2)$

from minutes upwards — 5. Solve the structure: (a) Patterson methods (b) Direct methods (c) Other methods — some or all non-H atom positions

minutes/hours — 6. Complete the structure — find all the atoms: Fourier and difference Fourier syntheses — all atom positions (approximate)

minutes—days — 7. Refine the structure model — atom positions and displacement parameters

? — 8. Interpret the results — molecular geometry, packing arrangement, etc....

Fig. 2.1 A flowchart for the steps involved in a crystal structure determination.

Figure 2.1 shows an outline of crystal structure determination in a simplified form as a schematic flowchart. The steps involved are in the boxes. To the right of each is listed the information obtained and to the left an indication of the time-scale involved in carrying out the operation. Some of these times vary considerably, depending on the quality of the sample being studied, the resources available for the work, the size and complexity of the structure, the skill of the crystallographer, and a certain amount of luck.

2.2   The preparation and selection of samples

The sample must be a *single crystal*, in which all the unit cells are identical and are aligned in the same orientation, so that they scatter cooperatively to give a clear diffraction pattern consisting of individual X-ray beams, each in a definite direction. Outward appearance such as regularity of shape is not important, but rather the internal regularity of the molecular arrangement on a well-defined lattice; many single crystals have an unpromising irregular shape, while polycrystalline and even non-crystalline materials such as glass may have beautiful external forms.

The intensities of X-rays diffracted by a crystal are proportional to its volume (more electrons give more scattering), but X-rays are also absorbed by crystals and this effect increases exponentially with crystal dimensions; absorption affects the measured intensities, introducing a systematic error, for which a correction may need to be made (see later). The amount of absorption depends on the X-ray wavelength and on the chemical composition, and can be very high when heavier elements are present. Systematic errors are also produced if the crystal is not completely bathed in the incident X-ray beam throughout the diffraction measurements, and most X-ray beams are less than 1 mm in cross-section. A typical acceptable crystal size is a few tenths of a millimetre; a smaller size and uniform dimensions are preferable for samples containing heavy atoms, and very small crystals can be examined with intense synchrotron radiation. Such crystals, much smaller than the popular image, usually need to be examined and handled under a microscope. A microscope with polarizing filters provides some useful optical tests of the quality of a crystal, but the ultimate test is the X-ray diffraction pattern. Crystals can be cut with a sharp scalpel, but this sometimes adversely affects the crystal quality.

Suitable crystals are sometimes produced in the initial synthesis of a compound, but often recrystallization is necessary. This process can be difficult, unpredictable, frustrating and time-consuming and is not guaranteed to succeed. The objective is quite different from that of recrystallization in synthesis—good quality single crystals of suitable size, with high yield not a priority—though both aim for a pure material, and special techniques have been devised. It is not uncommon for a crystal structure to incorporate molecules of solvent, so the solvent itself is one of the conditions which can be varied in the quest for suitable crystals.

One single crystal is separated from the rest of the sample and is mounted on a device which will hold it firmly in the X-ray beam; a precision of hundredths of a degree is required. Since the diffraction experiment involves rotating the crystal in the beam during exposure, as explained later, lateral adjustments need to be available to position the crystal accurately on each rotation axis. For some less commonly used equipment, it is an advantage if one unit cell axis can be aligned in a particular direction, so there may also be provision for angular adjustments. Such a device, known as a *goniometer head*, is shown in Fig. 2.3. Apart from the sample itself, no crystalline material should be in the X-ray beam, so the crystal is usually glued (with a minimum quantity of an amorphous glue) to a fine glass fibre attached to the goniometer

In reality, all crystals have faults in their internal structure, so unit cells are not exactly aligned. The range of misalignment is called the *mosaic spread*, because the slightly misaligned sub-microscopic blocks of a real single crystal resemble the tiles in a mosaic (Fig. 2.2). For a good quality single crystal, the mosaic spread is only a fraction of a degree.

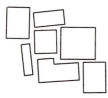

Fig. 2.2   Mosaic structure of a single crystal (highly exaggerated).

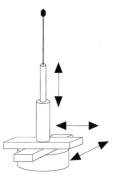

Fig. 2.3   Schematic representation of a *goniometer head* with a mounted crystal. Such devices were originally used on equipment for the optical measurement of angles between well developed flat faces of crystals, which is the derivation of the word goniometer (angle measuring device).

head (Fig. 2.4). The glue and glass contribute to general background scattering but not to the sharp diffraction maxima.

Samples which are air-sensitive or which degrade by loss of loosely bound solvent require special treatment. Handling them in an inert-atmosphere glovebox is possible but difficult. They may be sealed in thin-walled glass capillary tubes, an operation which is considerably easier if brief exposure to the air can be tolerated. Alternatively, the crystals can be coated with an inert viscous oil and then manipulated without difficulty under a normal microscope in the open atmosphere; if the X-ray examination is to be carried out at a sufficiently low temperature that the oil vitrifies to a glass, it can be used simultaneously as an adhesive and a protective coating, and this provides a particularly elegant and simple solution for materials of even extreme air sensitivity.

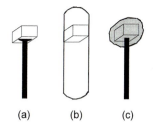

Fig. 2.4 Crystal mounting methods: (a) glued to a fine glass fibre; (b) enclosed in a capillary tube; (c) coated with an inert oil.

2.3 X-ray photography

As noted in Chapter 1, diffracted X-rays can be recorded on photographic film. The diffraction conditions represented by the Bragg equation are severe, and will be satisfied for only very few reflections for a randomly oriented stationary crystal in an X-ray beam, because few of the lattice planes will fortuitously be oriented at the correct θ angle, so the pattern recorded on film will show only a few spots (Fig. 2.5(a)).

In order to bring more lattice planes into a reflecting position, the crystal must be rotated in the X-ray beam (Fig. 2.5(b)). Recording the whole of the diffraction pattern on one film, however, leads to severe overlap of the reflections occurring at different stages of the rotation, because three-dimensional information is being compressed into a two-dimensional record, and its measurement and interpretation are impossible (Fig. 2.5(c)).

Instead, selected portions of the diffraction pattern need to be recorded separately on different films. The interpretation of them is greatly assisted if the rotation of the crystal is about the direction of a unit cell axis, and further simplification results from some types of correlated movement of the film with that of the crystal, and/or from the use of film held in a cylindrical mount rather than a flat one. Assigning *hkl* indices to the individual reflections is then

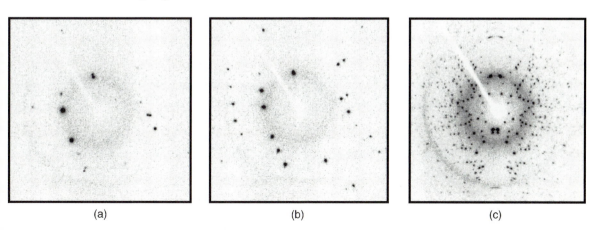

(a)	(b)	(c)

Fig. 2.5 (a) A diffraction pattern recorded on an electronic area detector from a stationary crystal; a similar pattern would be obtained on a photographic film, which is effectively a non-electronic area detector. (b) A diffraction pattern recorded from a 5° rotation of the same crystal. (c) A diffraction pattern recorded from a full 360° rotation of the same crystal.

An *X-ray camera* is an instrument for recording X-ray diffraction patterns on photographic film. A *diffractometer* is an instrument which records diffraction patterns by means of some kind of X-ray sensitive detector other than photographic film, usually involving the conversion of incident X-ray energy into an electronic signal.

a simple matter of counting along obvious rows of spots. Several types of *X-ray camera* have been developed over many years to achieve such effects, each operating with a particular geometrical combination of crystal orientation, film shape and film movement. The use of film methods for recording X-ray diffraction patterns has declined in recent decades as diffractometers have become more widespread, but they do offer the advantages of a permanent durable record of the whole diffraction pattern and its background in compensation for their slowness and non-digital output, which requires further optical measurement before it can be fed to a computer. The geometry, symmetry and intensities of the diffraction pattern can all be obtained from a suitable set of photographs, and many earlier crystal structures were determined in this way.

2.4   X-ray diffractometers

From around the 1960s, computer-controlled *diffractometers* became the standard means of collecting diffraction data. Instead of photographic film, an electronic device is used which is sensitive to X-rays. The most commonly used detector, a *scintillation counter*, contains a material, such as thallium-doped sodium iodide, which produces light in the visible region when X-rays fall on it. The light is detected and the signal amplified by a photomultiplier, so that the overall effect is an electrical pulse for each X-ray photon incident on the face of the detector, which is typically a few millimetres in diameter. In this way the intensities of individual reflections can be measured. For each one, the detector must be moved round one axis (usually vertical) to the correct 2θ angle. Because the detector can see only reflections which occur in the horizontal plane, more than one axis of rotation is needed for the crystal. The most widely used type of diffractometer has three rotation axes for the crystal, giving more than enough freedom, so that there is even a choice of settings possible for many of the reflections (Fig. 2.6). With one of these *four-circle diffractometers* reflections (positions and intensities) are observed one at a time, the crystal and detector being moved under computer control from each one to the next in sequence.

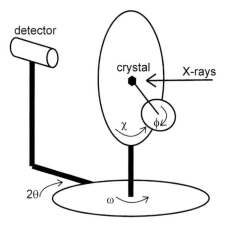

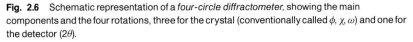

Fig. 2.6   Schematic representation of a *four-circle diffractometer*, showing the main components and the four rotations, three for the crystal (conventionally called ϕ, χ, ω) and one for the detector (2θ).

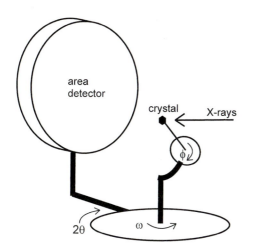

Fig. 2.7   Schematic representation of a diffractometer equipped with an *area detector.*

More recently, X-ray detectors have become available which record over a considerably larger area and are position-sensitive: a number of incident beams can be recorded at the same time, and their positions as well as intensities are known. There are various types of *area detectors* based on different technologies, each with particular advantages and disadvantages of size, sensitivity, spatial resolution, speed of read-out and cost, and this is a topic of considerable current development; they can be regarded simply as electronic equivalents of photographic film in many respects. An area detector can replace the scintillation counter of a four-circle diffractometer, but it is also possible to reduce the number of rotation axes for the crystal, because of the size of the detector (Fig. 2.7); it is no longer necessary to bring all reflections into the horizontal plane in order to record them.

After many years of relative stability in the design of diffractometers, the introduction of area detectors has brought considerable change and opened up exciting new possibilities (see examples later).

2.5   Obtaining unit cell geometry and symmetry

Both photographically and with a diffractometer the unit cell geometry can be measured from a subset of the complete diffraction pattern. The key step is assigning the correct indices *hkl* to each of the observed reflections. From these and the measured Bragg angle for a few reflections, the six unit cell parameters can be calculated via the Bragg equation and modified versions of it appropriate to the geometry of the particular camera or diffractometer being used.

With photographs, one unit cell axis is aligned in a particular direction relative to the incident X-ray beam and this produces a regular pattern of spots on the film, to which indices can easily be assigned by counting along rows. Several photographs are required to give the full three-dimensional geometry, and each may take several hours.

With diffractometers, the crystal is usually mounted in a random orientation, and this has to be determined as well as the unit cell geometry. Some reflections of moderate to high intensity are located by simply driving the various motors while monitoring the detector output for a signal significantly above background (a blind search, all under computer control), or by taking short exposure Polaroid photographs and measuring the positions of spots on them. From these positions, perhaps one or two dozen of them for a four-circle diffractometer, rather more for an area detector, the crystal orientation, unit cell geometry and reflection indices have to be determined simultaneously, by calculations which are not simple and are usually regarded as computer 'black-box' methods, but they are all based essentially on the Bragg equation. A diffractometer will often give a unit cell and orientation for a crystal in less than an hour; a few minutes are usually enough with an area detector.

At this stage, it may be possible to assign the correct space group by comparison of intensities which are equivalent by symmetry, and by noting that certain special subsets of reflections have zero intensity (Fig. 2.8), which is an effect of symmetry elements with a translation component (glide planes and screw axes), but the decision is made more reliably on the basis of the complete data set later. We have already seen examples of how this may provide some information about the structure, such as molecular symmetry or presence of solvent.

Of course, the initial examination of a crystal with X-rays also shows the quality of the diffraction pattern, from which a decision is made whether to proceed with the full experiment or look for a better crystal.

2.6   The measurement of intensities

Although diffraction intensities can be measured from photographic films, this is now rarely done. It involves estimating the degree of blackening in each spot, which can be achieved either by visual comparison with a calibrated scale or by measuring the absorption of a beam of light passed through the film. Some reflections are too weak to be seen above the general level of background scattering on the film, and these are labelled as 'unobserved'; usually no numerical value is recorded for their intensities, and they are not used in the successive calculations. The process of estimating intensities may take several weeks, depending on the size and symmetry of the structure (and hence the number of photographs required to record the complete data set) and the overall intensity of diffraction.

A conventional four-circle diffractometer measures intensities one at a time in an automatic, computer-controlled process. For each reflection the crystal and detector are driven to the appropriate positions to satisfy the Bragg equation and bring the diffracted beam into the detector in the horizontal plane, and the total 'integrated intensity' is measured while the crystal is rotated through a small angle from one side of the Bragg position to the other to allow for the mosaic spread of the crystal, which produces a peak profile of a few tenths of a degree rather than a sharp spike of intensity at one angle (Fig. 2.9). Some diffractometer systems carry out a detailed statistical analysis of the reflection profile shape, which provides more reliable results for weaker reflections.

The crystallographer must make some decisions about the data collection procedure. These include the maximum Bragg angle to be measured (reflections at higher angles are generally weaker but add to the precision of

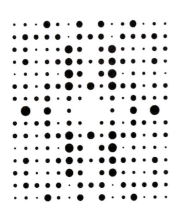

Fig. 2.8 Part of a diffraction pattern showing systematic absences (alternate reflections missing on the central horizontal and vertical rows) due to screw axes.

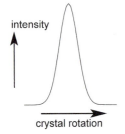

Fig. 2.9 A reflection profile observed on a diffractometer as the crystal is rotated.

the final structure if they have measurable intensities), the time to spend on each measurement, and whether to measure only the *unique set* of data (a fraction of the total pattern depending on the space-group symmetry) or to include other reflections equivalent by symmetry, which takes longer but again improves the quality of data overall and gives a confirmation of the symmetry. The time taken to collect the intensity data on a four-circle diffractometer depends very much on these decisions and on the size of the structure; a larger structure gives more reflections to the same maximum Bragg angle. It takes at least overnight and usually several days.

With an area detector diffractometer, many diffracted beams are recorded simultaneously. Usually the crystal is rotated about one axis and each exposure covers a small angular range, the details depending on the particular characteristics of the detector: those with rapid read-out of the image can efficiently measure diffraction patterns with relatively narrow angular slices, but slower read-out demands fewer, wider angular frames. In general, reflections are spread over more than one frame and sophisticated computer analysis of large quantities of data is required. Among other advantages, area detectors can provide a high degree of redundancy of symmetry-equivalent data and of the same reflections measured more than once. Data collection typically takes only a few hours, independent of the size of the structure, since a larger structure just gives more simultaneous reflections, but longer exposures are advisable for weakly scattering samples. Here the use of synchrotron radiation brings a substantial improvement.

The result of this process, from whatever equipment is used, is a list of reflections, usually thousands of them, each with *hkl* indices and a measured intensity. In addition, from diffractometer measurements, each intensity I has an associated *standard uncertainty* (s.u.), $\sigma(I)$, which is calculated from the known statistical properties of the X-ray generation and diffraction processes, and is a measure of the precision or reliability of the measurement.

The *unique set* of data, up to a particular maximum Bragg angle, is the total set of reflections which are independent of each other by symmetry. Application of symmetry to the unique set produces all the reflections which can be measured. For example, in the centrosymmetric triclinic case the diffraction pattern has only inversion symmetry, and the unique set is exactly half of the total available data: each reflection h, k, l is equivalent to the reflection $h, -k, -l$. For a centrosymmetric monoclinic structure, the unique set is one quarter of the total, and for centrosymmetric orthorhombic it is one eighth.

A previous term for standard uncertainty, still in wide use, is *estimated standard deviation* (e.s.d.). It is an estimate of the spread of values which would be obtained if the measurements were repeated many times.

2.7   Data reduction

We have previously seen that the intensity of an X-ray beam is proportional to the square of the wave amplitude. The measured intensity is affected by various factors, however, for which corrections must be applied. The conversion of intensities I to 'observed structure amplitudes' $|F_o|$ (o = observed) or F_o^2 and, correspondingly, of s.u.'s $\sigma(I)$ to $\sigma(F_o)$ or $\sigma(F_o^2)$ is known as data reduction and has several components.

$I(hkl) \propto |F(hkl)|^2$

There are corrections associated with the data collection process, which are geometrical in nature. These are a function of the geometry of the equipment and so are instrument-dependent. There is also a correction needed because reflected radiation is partially polarized (a phenomenon exploited in the use of polaroid sunglasses, for example). These geometrical corrections, known as *Lorenz-polarization* factors, are well known and easily made.

A correction may also be needed for changes in the incident X-ray beam intensity or in the scattering power of the crystal during the experiment. The former is important for synchrotron radiation, which decays gradually and significantly and can be directly monitored, and the latter may be caused by some decomposition of the sample in the high-energy X-ray beam. The effect of both is to make intensities measured later in the experiment weaker than they

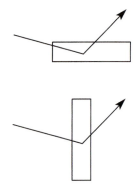

Fig. 2.10   The effect of absorption for a needle-shaped crystal.

In reality, of course, intensities cannot be negative, but the weakest reflections from a diffractometer may be insignificantly above background and, through statistical variations in the measuring process, may be recorded as below background, i.e. apparently net negative. These would be 'unobserved' by photographic methods. The fact that they are weak is actually valuable information in structure determination.

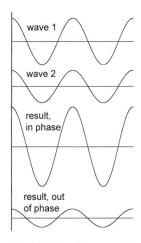

Fig. 2.11   Addition of two waves (first and second) to give their sum (in phase, third) or their difference (out of phase, fourth).

should be relative to earlier ones. A correction can be made on the basis of reflections which are measured repeatedly at intervals during the data collection.

Where absorption effects are significant a further correction must be made. Each reflection is affected differently by absorption, because the absorption depends on the path length of the X-rays through the crystal, and this varies as the crystal orientation is changed (Fig. 2.10). Many different types of absorption correction are used. Some are based on careful measurement of the crystal shape and dimensions and calculation of path lengths; others are based on comparison of intensities of symmetry-equivalent reflections, which should be equal but are not because of absorption effects.

The data reduction process also includes the merging and averaging of repeated and symmetry-equivalent measurements in order to produce a unique, corrected and scaled set of data. This calculation affords a numerical measure of the agreement among equivalent reflections, which is one indication of the quality of the data and the appropriateness of the applied corrections.

At the same time, statistical analysis of the complete unique data set can provide an indication of the presence or absence of some symmetry elements, particularly whether the structure is centrosymmetric or not, though this is not infallible; and the observed overall decay of intensity with increasing $(\sin\theta)/\lambda$ gives an average atomic displacement parameter.

The various corrections for the intensities are applied also to their s.u.'s. The result of this whole process, which usually takes only a matter of minutes on a computer, is a list of reflections as $h, k, l, |F_o|, \sigma(F_o)$ [or $h, k, l, F_o^2, \sigma(F_o^2)$; the advantage of retaining the squared form is that no special treatment is required for intensities measured as negative].

2.8   Solving the structure

Having measured and appropriately corrected the diffraction data, we turn now to the solution of the structure, in which we obtain atomic positions in the unit cell from the data. Remember that the objective here is to imitate a microscope lens system, recombining the individual diffracted beams to give a picture of the electron density distribution in the unit cell.

$$\rho(xyz) = \frac{1}{V} \sum_{h,k,l} |F(hkl)| \cdot \exp[i\phi(hkl)] \cdot \exp[-2\pi i(hx + ky + lz)] \quad (1.18)$$

The mathematical expression of this process is equation 1.18, repeated above. The amplitudes $|F(hkl)|$ have been measured, the final exponential term can be calculated for the contribution of each reflection hkl to each position xyz, but the phases of the reflections are unfortunately unknown, so the calculation cannot be carried out immediately.

The result of adding even just two waves varies from the sum to the difference of their amplitudes depending on their relative phases (Fig. 2.11), and to apply trial-and-error methods to thousands of waves is a task of impossible proportions.

We shall see in the next step (section 2.9) that knowing part of the structure, i.e. the positions of some of the atoms, especially those with the most electrons, is often enough to help find the rest. The question is, where to begin?

Of the various methods used, two are by far the most common and important. One works best for structures containing one atom or a small number of atoms with significantly more electrons than the rest ('heavy atoms'), while the other is more appropriate for 'equal atom' structures. In general, not surprisingly, the easiest atoms to find are those which contribute most to the total scattering.

The Patterson synthesis

The Fourier transform of the observed diffracted beam amplitudes $|F_o|$ gives the correct electron density, but it requires knowledge of the phases of all the reflections. The Fourier transform of the squared amplitudes F_o^2 with all phases set equal to zero (all waves taken in phase) produces what is called a *Patterson synthesis* (or Patterson function, or Patterson map).

$$P(xyz) = \frac{1}{V} \sum_{h,k,l} |F_o(hkl)|^2 \cdot \exp[-2\pi i(hx + ky + lz)] \qquad (2.1)$$

A. L. Patterson introduced and developed this method relatively early in the history of X-ray crystallography.

All the information needed for this transform is known; it can be calculated for any measured diffraction pattern. But is it of any use?

The Patterson map looks rather like an electron density map (see Fig. 2.12), in that it has peaks of positive density in various positions. These are not, however, the positions of atoms in the structure. Instead, it turns out that the Patterson function is a map of vectors between pairs of atoms in the structure. For every pair of atoms at positions (x_1, y_1, z_1) and (x_2, y_2, z_2) there is a peak in the Patterson map at $(x_1 - x_2, y_1 - y_2, z_1 - z_2)$ and another of the same size at $(x_2 - x_1, y_2 - y_1, z_2 - z_1)$. In other words, for every peak seen in the Patterson map (at, say, u, v, w), there must be two atoms in the structure whose x coordinates differ by u, y coordinates differ by v, and z coordinates differ by w. The Patterson peaks show where atoms lie relative to each other, but not where they lie relative to the unit cell origin, which is what we really want to know.

Peaks in an electron density map are, ignoring vibration effects, proportional in size to the atomic numbers of the respective elements, since these are equal to the number of electrons. Patterson peaks are proportional in size to the product of the atomic numbers of the two atoms concerned.

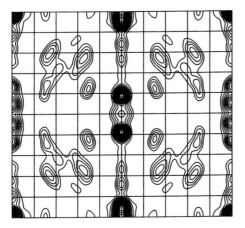

Fig. 2.12 A section of the Patterson map for a structure containing one unique As atom together with atoms of H, C, N, F and S.

In order to see how the Patterson function can be used to locate some of the atoms, we note some of the properties of the Patterson function, which follow from its definition.

(a)   Every atom forms a pair, and hence a vector, with every other atom, including with itself. So a unit cell containing n atoms gives n^2 vectors. Of these, the n self-vectors (each atom to itself) have zero length and all coincide at the origin $(0, 0, 0)$. This is always the largest peak in any Patterson map. There are $n^2 - n$ other peaks.

(b)   The vector between atom A and atom B is exactly equal and opposite to the vector between atom B and atom A. This means that a Patterson map always has an inversion centre, even if the crystal structure itself does not.

(c)   Patterson peaks have a similar shape to electron density peaks, but are about twice as broad.

(d)   As a consequence of points (a) and (c), there is usually considerable overlap of peaks, and not all will be resolved as separate identifiable maxima.

For these reasons, Patterson maps usually show large featureless regions of overlapped broad peaks, with significant peaks due to vectors involving 'heavy atoms'. If a structure contains only a few heavy atoms among a lot of lighter atoms, the Patterson map will show a small number of large peaks standing out clearly above the general background level.

In such cases it is usually possible to find a self-consistent set of atomic positions for the heavy atoms which explain the large Patterson peaks. Vectors between symmetry-related heavy atoms often lie in special positions, with some coordinates equal to 0 or $\frac{1}{2}$, for example, and are easily recognized. Examples are given in the next chapter. Solving a Patterson map is rather like a mathematical brain-teaser puzzle.

Once the heaviest atoms have been found, the rest are located as shown in section 2.9.

For example, for ferrocene $Fe(C_5H_5)_2$, Fe–Fe peaks, Fe–C peaks and C–C peaks have relative heights $26 \times 26 = 676$, $26 \times 6 = 156$, and $6 \times 6 = 36$; these are in the ratio approximately $19 : 4 : 1$. Peaks involving H are even smaller. The few large Fe–Fe peaks (more than one molecule per unit cell!) will be clearly seen among the many smaller overlapped peaks. Figure 2.14 also shows a small number of large peaks and is otherwise relatively featureless.

Patterson search methods

Even for structures without particularly heavy atoms, the Patterson synthesis can provide a solution method in some cases. If a significant proportion of the molecule has a known shape, then a group of vectors generated internally by these atoms can be calculated. Such a pattern occurs in the Patterson map, but its orientation is unknown and it is mixed up with other vectors involving the rest of the molecule and vectors between atoms in different molecules. It may be possible to match the pattern by computer search and hence find the correct orientation of the known fragment. Further computer analysis of possible intermolecular vectors then gives proposed positions for the fragment. Steroids, with a relatively rigid and predictable tetracyclic nucleus, are good examples of suitable materials for this approach (Fig. 2.13). *Direct methods* are, however, much more commonly used for 'equal atom' structures.

Direct methods

This is a general name given to methods which seek to obtain approximate reflection phases from the measured intensities with no other information available.

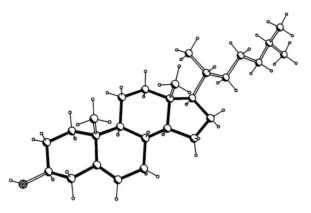

Fig. 2.13 The structure of cholesterol, with the characteristic rigid steroid tetracyclic framework highlighted.

Such a description of the situation is, however, misleading. There is other information available to help us find the missing phases, in the nature of the electron density we are trying to determine.

The electron density is the Fourier transform of the diffraction pattern. This means we add together a set of waves in order to produce the electron density distribution. Each wave has half its value positive and half negative (alternate 'crests' and 'troughs': see Fig. 1.3), except for $F(000)$, which is constant and positive. The electron density, however, is everywhere positive or zero; it can have no negative regions. Furthermore, it is concentrated into certain compact regions (atoms; see Fig. 1.5). So the waves must be added together in such a way as to build up and concentrate positive regions and cancel out negative regions. This puts considerable restrictions on the relationships among the phases of different reflections, especially the most intense ones, which contribute most to the sum.

Since large numbers of reflections are involved in the complete Fourier transform, individual phase relationships are not certainties, but have to be expressed in terms of probabilities, and the probabilities depend on the relative intensities.

Direct methods involve selecting the most important reflections (those which contribute most to the Fourier transform), working out the probable relationships among their phases, then trying different possible phases to see how well the probability relationships are satisfied. For the most promising combinations (assessed by various numerical measures), Fourier transforms are calculated from the observed amplitudes and trial phases, and are examined for recognizable molecular features.

Over the years, various methods of more or less sophistication have been developed for the steps involved. They can be regarded most simply as a sort of inspired trial-and-error method, in which it is usually necessary to try many different sets of phases and use the relationships themselves to 'refine' or improve them. Direct methods involve a considerable amount of computing, and are treated as a 'black box' even by many of their regular users. When they are successful, they usually locate most or all of the non-hydrogen atoms in a structure. Examples are given in the next chapter, and an illustration in one dimension is shown below.

Other methods

Almost all 'small molecule' crystal structures these days are solved by either Patterson or direct methods. Lack of space precludes a discussion of other methods used for macromolecular structures, which rarely contain heavy atoms with a high proportion of the total electron density, and for which the very large number of reflections and hence of phase relationships makes conventional direct methods less reliable.

A partial solution for some small molecule structures can be found from considerations of symmetry. For example, if a molecule which could reasonably have a centrosymmetric geometry crystallizes in the triclinic system with one molecule per unit cell, then the centre of the molecule probably coincides with a crystallographic inversion centre in space group $P\bar{1}$ (see section 1.5). Such a situation is frequently found for metal complexes, and the metal atom is thereby located at a special position. If the metal has sufficient electrons to be classed as a heavy atom, it can be used, as shown in the next section, to find the rest of the atoms; no Patterson map or direct methods calculation is necessary.

It should be noted that there is no 'correct' method for solving a particular structure. Once the right solution has been found, by whatever method, it can be further refined (see below); the method of solution is no longer important. If one method does not work, it is perfectly valid to try others, with all the available variations, until one is successful. The objective is to beat the 'phase problem'; exactly how this is done does not really matter.

A one-dimensional illustration of direct methods and Patterson synthesis

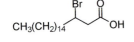

Fig. 2.14 The formula of 3-bromo-octadecanoic acid.

Racemic 3-bromo-octadecanoic acid forms triclinic crystals with two molecules in the unit cell, related to each other by inversion symmetry. The unit cell has two short and one long axes, and the molecule is extended approximately along the longest axis (*c*). This structure provides an illustration of structure solution in one dimension; if we take only those reflections (00*l*) which have *h* and *k* indices equal to zero, they contain no information about the *x* and *y* coordinates of atoms, but we can use them to find *z* coordinates. The reflections (001) and (002) were not measured, probably because they lie too close to the direct beam; intensities were obtained for reflections with *l* from 3 to 21, and their amplitudes, derived from the measured intensities, are in Table 2.1.

To find the *z* coordinates of the atoms, it is necessary to add up the contributions of all these 19 waves, together with $F(000)$, at each of a range of *z* values from 0 to 1 (because of the inversion symmetry in the structure, the

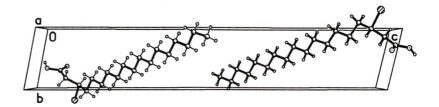

Fig. 2.15 Two molecules in the elongated unit cell of 3-bromo-octadecanoic acid.

range 0.5 to 1.0 is actually equivalent to the range 0.0 to 0.5 by inversion, but we shall carry out the calculations for the whole z range of one unit cell for completeness). For this purpose we need a one-dimensional version of equation 1.18.

$$\rho(z) = \frac{1}{c}\sum_{l}|F(l)| \cdot \exp[i\phi(l)] \cdot \exp[-2\pi i(lz)] \quad (2.2)$$

At each chosen value of z, there are 19 terms to add together. The task is further simplified by the fact that the structure is centrosymmetric (this simplification for centrosymmetric structures applies also in three dimensions). In such cases the phases of reflections can only take one of two special values, 0 and 180° (or 0 and π radians), and the term $\exp(i\phi)$ becomes equal to $\cos(\phi)$; this is simply $+1$ for $\phi = 0$, and -1 for $\phi = 180°$, and the mystery of the unknown phase narrows down to a straight choice between completely in phase and completely out of phase for each individual reflection. This is the same as having to choose whether each amplitude is added or subtracted to make up the total sum. This still leaves us with a two-way choice 19 times over, giving 524 288 possible combinations! This is clearly too much for a blind trial-and-error approach.

Note also that the final exponential term simplifies to a cosine in the same way, so we have the equation

$$\rho(z) = \frac{1}{c}\sum_{l}|F(l)| \cdot \text{sign}[F(l)] \cdot \cos[2\pi(lz)] \quad (2.3)$$

For the one-dimensional case, this summation process can be shown graphically. The contributions of each of the 19 reflection amplitudes to each point across the range of z from 0 to 1 are shown in Fig. 2.16. The contributions are all shown on the same scale, and with all the unknown signs (corresponding to the unknown phases) set as positive. The cosine term in equation 2.3 means that reflections with a low value of the index l make a contribution to the electron density which varies only slowly across the unit cell; reflections with a high value of l contribute much more finely, with more maxima and minima across the range. Figure 2.17 shows the result of adding up these 19 contributions with different combinations of signs (phases). The top result comes from arbitrarily chosen signs; it does not look like a promising solution for the electron density, particularly with some deep minima. The second result comes from the correct signs (as provided by a forward Fourier transform calculation from the final known structure; these correct signs are given in Table 2.1); it shows large maxima for two symmetry-related bromine atoms (two molecules per unit cell) and smaller maxima, approximately regularly spaced, most of which correspond to pairs of carbon atoms, these overlapping in projection along the axis. The positions of the atoms, particularly the large bromine atoms, are correctly given: the final refined z coordinates for bromine are very close to 0.1 and 0.9.

To illustrate the principles of direct methods, concentrate on the largest amplitudes; clearly these contribute most to the summations, and incorrect signs for the smaller amplitudes will not greatly affect the result. The correct signs for reflections 4 and 5 (using the l indices to label them) are both negative. This means they should both be turned upside down before being added into the sum. Together they then contribute a considerable positive

Table 2.1 Observed amplitudes and correct phases for 00*l* reflections of 3-bromo-octadecanoic acid

l index	Measured $|F(00l)|$	Correct sign
3	5.8	+
4	45.2	−
5	39.2	−
6	52.6	−
7	10.6	−
8	3.8	+
9	32.2	+
10	31.8	+
11	30.4	+
12	11.8	+
13	6.2	−
14	18.2	−
15	21.8	−
16	16.2	−
17	8.2	−
18	10.0	+
19	14.4	+
20	23.4	+
21	44.6	+

index	sign	|F|
3	+	5.8
4	−	45.2
5	−	39.2
6	−	52.6
7	−	10.6
8	+	3.8
9	+	32.2
10	+	31.8
11	+	30.4
12	+	11.8
13	−	6.2
14	−	18.2
15	−	21.8
16	−	16.2
17	−	8.2
18	+	10.0
19	+	14.4
20	+	23.4
21	+	44.6

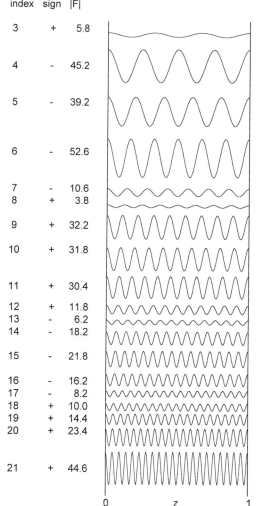

Fig. 2.16 The contributions of the 19 00*l* reflections to the one-dimensional Fourier summation of equation 2.3, with all their phases set at zero. The correct phases, as signs, are shown with the indices and amplitudes in the left-hand columns. Reflections with larger indices are observed at higher Bragg angles and provide greater resolution of the electron density image, just as light scattered at greater angles by an object on an optical microscope provides better resolution than low-angle scattering. The curves shown here for the different reflections must not be confused with X-ray wavelengths and frequencies (the wavelength is constant for all reflections); these are not the waves themselves, but the contributions they make to the electron density calculation at various points in the unit cell via the one-dimensional Fourier transformation.

amount where their original first and last troughs, as seen in Fig. 2.16, almost coincide, a large negative amount at $z = 0$ and $z = 1$, and their contributions largely cancel out elsewhere. Reflection 9 is also quite strong, and the positions of its crests and troughs are related to those of reflections 4 and 5, simply because $4 + 5 = 9$. If reflection 9 is to reinforce the positive build-up of electron density provided by the inverted reflections 4 and 5, rather than partially cancelling it, it must have a crest roughly coinciding with the first troughs of those reflections as they are shown in the figure, so its sign must be positive and not negative. Since $(-1) \times (-1) = (+1)$ this relationship can be represented as

$$\text{sign}[F(9)] = \text{sign}[F(4)] \times \text{sign}[F(5)] \tag{2.4}$$

or, in terms of phase angles rather than signs,

$$\phi[(9)] = \phi[(4)] + \phi[(5)] \tag{2.5}$$

The probability that such a relationship among the phases of reflections with related indices is true increases with their amplitudes. In this particular example, if we take all the reflections with amplitudes greater than 80, there are 11 of these relatively strong reflections and 19 relationships of this kind which involve only these reflections. All 19 relationships are, in fact, obeyed [for example, reflection 6 is negative, reflection 9 is positive, and reflection 15 is negative: $(-1) \times (+1) = (-1)$]; Table 2.2 gives them all. In general, for three-dimensional structures, because these relationships are probabilities rather than certainties, some of the indications will be wrong, and for some reflections participating in several relationships contrary indications may be found, so an overall balance of the various indications has to be taken.

Table 2.2   Phase relationships for the strongest 00*l* reflections

	<u>4</u>	<u>5</u>	<u>6</u>	9	10	11	<u>14</u>	<u>15</u>	<u>16</u>	20	21
<u>4</u>		9	10		<u>14</u>	<u>15</u>			20		
<u>5</u>	9	10	11	<u>14</u>	<u>15</u>	<u>16</u>		20	21		
<u>6</u>	10	11		<u>15</u>	<u>16</u>		20	21			
9		<u>14</u>	<u>15</u>			20					
10	<u>14</u>	<u>15</u>	<u>16</u>		20	21					
11	<u>15</u>	<u>16</u>		20	21						
<u>14</u>			20								
<u>15</u>		20	21								
<u>16</u>	20	21									
20											
21											

For each relationship the *l* indices of the three reflections are given by one entry in the table body together with the corresponding number at the head of the column and the number at the left-hand end of the row; an underlined index represents a negative reflection amplitude (for example, reflection 6 has a negative amplitude, but reflection 9 has a positive amplitude).

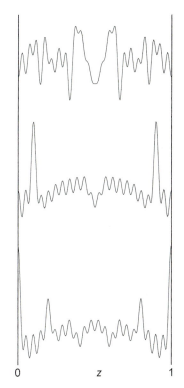

Fig. 2.17   Combinations of the 19 contributions of Fig. 2.16 with different sets of phases: top, randomly chosen phases giving an unrecognizable result; middle, correct phases clearly showing the bromine atoms; bottom, all phases positive, resembling a Patterson synthesis.

In direct methods as commonly applied, the strongest reflections are chosen and all the phase relationships among them are generated. Then various possible combinations of phases are tried, either by assigning values to a few reflections and using the probability relationships to generate others, or by assigning random phases to all the reflections and using the relationships to improve them so that they fit the relationships better. The combination of phases giving the best agreement with the expected relationships is then used, together with the observed amplitudes, in a reverse Fourier transformation to calculate an electron density map, and this is examined for recognizable molecular features. In typical cases a few tens of initial phase sets are tried and

It is not actually the strongest reflections absolutely which are most important, but those reflections which are relatively strong for their Bragg angle; because intensities always decrease at higher angle, a high-angle reflection which is weak compared with low-angle reflections but strong compared with other high-angle reflections is an important contributor to the electron density calculation. For this reason, direct methods use *normalized structure amplitudes*, more commonly known as *E values*, instead of the basic amplitudes (*F values*) themselves; normalization is a calculation which effectively compensates for the θ-dependent decrease in intensity within a data set.

several of these are likely to lead to a recognizable correct structure showing most or all of the non-hydrogen atoms. In more difficult cases, many hundreds or thousands of attempts may be necessary.

The third result in Fig. 2.17 comes from taking all the reflections as positive, i.e. all phases set at zero; the 19 curves of Fig. 2.16 are simply added together as they are. Since all the curves have a crest at $z = 0$ and at $z = 1$, very large maxima are generated at these points. This is just like a one-dimensional Patterson synthesis, except that $|F|$ values have been used instead of F^2 values. The corresponding result using F^2 and zero phases looks very much like the third curve except that the maxima and minima are more exaggerated. The large maximum at the origin is a feature of all Patterson syntheses, corresponding to the superposition of the self-vectors of all atoms in the structure. The next largest maximum in this curve (together with its symmetry equivalent) corresponds to a vector between the two bromine atoms in the unit cell; its z coordinate is twice the z coordinate for one bromine atom, because the two symmetry-equivalent bromine atoms are at $+z$ and $-z$, and the difference between these is $(+z) - (-z) = 2z$. Hence the bromine atom in the molecule can be located simply by inspection of this Patterson synthesis, without the knowledge or guessing of any phases at all. Location of the remaining atoms then follows in the next stage (section 2.9). The solution of a Patterson synthesis in order to find one unique heavy atom in a structure (together with its symmetry equivalents) often involves no arithmetic more difficult than dividing by 2. Further examples are given in Chapter 3.

2.9   Completing the basic structure

If the initial structure solution has revealed positions for all the atoms (except for hydrogen atoms, which have very little electron density and are not usually found until later, if at all), this next step is unnecessary. Often, however, particularly from analysis of a Patterson map, only a partial structure has been obtained: some atom positions are known, but not all. This partial structure serves as our initial model or trial structure.

Using the forward Fourier transform equation (the mathematical representation of the scattering process), we can calculate what the diffraction pattern would be if this model structure were, in fact, the correct complete structure:

$$\text{model structure} \xrightarrow{\text{FT}} \text{set of } F_c \tag{2.5}$$

where F_c are calculated structure factors, one corresponding to each observed structure factor F_o. The calculation provides values for the amplitudes and phases of F_c ($|F_c|$ and ϕ_c), whereas we have only amplitudes for F_o ($|F_o|$, no ϕ_o).

If the atoms of the model structure are approximately in the right positions, there should be at least some degree of resemblance between the calculated diffraction pattern and the observed one, i.e. between the sets of $|F_c|$ and $|F_o|$ values. The two sets of values can be compared in various ways. The most widely used assessment is a so-called *residual factor* or *R-factor*, defined as

$$R = \frac{\sum ||F_o| - |F_c||}{\sum |F_o|} \tag{2.6}$$

This involves adding together all the discrepancies between corresponding observed and calculated amplitudes, ignoring signs of the differences, and normalizing the sum by dividing by the sum of all the observed amplitudes to give a value which can be compared for different structures. Variations on this definition include using F^2 values instead of $|F|$ values, squaring the differences, and/or incorporating different weighting factors multiplying different reflections, based on their s.u.'s, and hence incorporating information on the relative reliability of different measurements; for example, one residual factor in a very widely used computer program for crystal structure determination is

$$wR2 = \sqrt{\frac{\sum w(F_o^2 - F_c^2)^2}{\sum w(F_o^2)^2}} \qquad (2.7)$$

wR2 is the conventional name for this residual factor. The w indicates that weights are included, and the 2 indicates that F^2 values are used rather than F values.

where each reflection has its own weight w. This is, in many ways, and certainly from a statistical viewpoint, more meaningful than the basic R factor. For a correct and complete crystal structure determined from well measured data, R is typically around 0.02–0.07; for an initial model structure it will be much higher, possibly 0.4–0.5 depending on the fraction of electron density so far found, and its decrease during the next stages is a measure of progress. Values of $wR2$ and other residual factors based on F^2 are generally higher than those based on F values, by a factor of two or more.

Obviously, the reverse Fourier transform (the mathematical representation of the image construction) carried out with the calculated amplitudes $|F_c|$ and calculated phases ϕ_c would just regenerate the electron density of the model structure

$$|F_c| \text{ with } \phi_c \xrightarrow{\text{FT}^{-1}} \rho \text{ for model structure} \qquad (2.8)$$

and this is not progress. However, combination of the experimentally observed amplitudes $|F_o|$ (which carry information about the true structure) with the calculated phases ϕ_c (which are not completely correct, but are the best approximation we currently have to the unavailable ϕ_o values) produces something new:

$$|F_c| \text{ with } \phi_c \xrightarrow{\text{FT}^{-1}} \rho \text{ for a new model structure} \qquad (2.9)$$

Usually, if the errors in the calculated phases are not too large, this electron density shows the atoms of the existing model structure, together with additional atoms not already known. This provides an improved model structure, with more atoms than before.

If there are still more atoms to be found, this process can be repeated. A forward Fourier transform of the new model structure gives a new set of $|F_c|$ and ϕ_c; the previous set is discarded. The new $|F_c|$ and the unchanged original $|F_o|$ values should now give a lower R-factor, and the improved ϕ_c together with $|F_o|$ generate, via another reverse Fourier transform, a further electron density map.

Eventually all the atoms are located and the Fourier transform calculations give no further improvement. This repeated process is an example of a *bootstrap* procedure.

There are some variations on the basic Fourier bootstrap procedure, which make it more effective. In particular, the reverse transform (Fourier map calculation) can be carried out using the differences $|F_o| - |F_c|$ instead of just

This is a term common in computing jargon, usually shortened to 'boot', and is derived from an old saying, 'You can't pull yourself up by your own bootstraps' i.e. shoe-laces; in computer operating systems, crystallography and many other sciences, you can!

$|F_o|$. In this case, a *difference electron density map* is produced, in which the existing atoms of the current model structure do not appear. This makes new atoms stand out more clearly from the background and from false maxima arising from errors in the ϕ_c values. Difference electron density peaks or holes (negative peaks) at model structure atom positions may indicate incorrect atom assignments with too little or too much assumed electron density, which should be corrected in the next model.

Full electron density maps, with values of electron density or difference electron density at each of many regularly spaced points in all or part of the unit cell, either numerically or as contours, are not often generated and output as such by computer programs used for chemical crystallography. In most cases an automatic search is carried out for the positions of maximum density (peaks, analogous to mountains on a geographical map), and the output is just a list of these in descending order of height, as potential positions of atoms.

An example of structure completion by Fourier syntheses

To illustrate the bootstrap procedure we take a structure which has most of the atoms lying in one plane, because the gradual development of the structure is clearly seen by successive calculations of the electron density in this plane. The compound is $[PhSNSNSNSPh]^+[AsF_6]^-$ (Fig. 2.18). The complete cation, together with arsenic and two flourine atoms of the anion, lie on a mirror plane perpendicular to the *b* axis in an orthorhombic space group, so all their *y* coordinates are equal. The position of the arsenic atom within this plane can be found by inspection of the Patterson synthesis, part of which was shown in Fig. 2.12 (As has 34 electrons, S 16, F 9, N 7 and C 6, so vectors between symmetry-equivalent arsenic atoms stand out clearly).

Calculation of the diffraction pattern corresponding to the arsenic atom alone (and its equivalent atoms according to the space group symmetry), by the forward Fourier transform equation (2.5), gives a set of calculated amplitudes and phases for this very crude initial model structure; at this stage the value of *R* is 0.604, so the agreement between the observed and calculated amplitudes is not very good. The calculated phases are also far from perfect (the structure is centrosymmetric, so each phase must be either 0 or 180°, and any particular calculated phase is either completely right or completely wrong), but there are enough correct phases for the reverse Fourier transform, calculated from these phases and the observed amplitudes (equation 2.9), to show not only the known arsenic atom, but also four clear peaks for the four sulfur atoms (Fig. 2.19(a)). Inclusion of these in the model structure leads to a calculated diffraction pattern in better agreement with the observed amplitudes: *R* is reduced to 0.364, the calculated phases are more nearly correct, and the next electron density map (reverse Fourier transform) shows all the N, C and F atoms (Fig. 2.19(b)); four of the F atoms are not seen in this figure, because they lie out of the plane shown, but peaks are found for them in the correct positions above and below the plane. With all the non-hydrogen atoms included, *R* drops to 0.036 after refinement (see below), and an electron density map shows well resolved and clear peaks for all the atoms (Fig. 2.19(c)). The largest peaks in a difference electron density map (calculated from $|F_o| - |F_c|$ instead of from $|F_o|$) are in the positions expected for the hydrogen atoms (Fig. 2.19(d)), and incorporation of these into the model structure leads to a final *R* of 0.027, and a very precise structure.

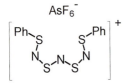

Fig. 2.18 The component ions of $[PhSNSNSNSPh]^+[AsF_6]^-$.

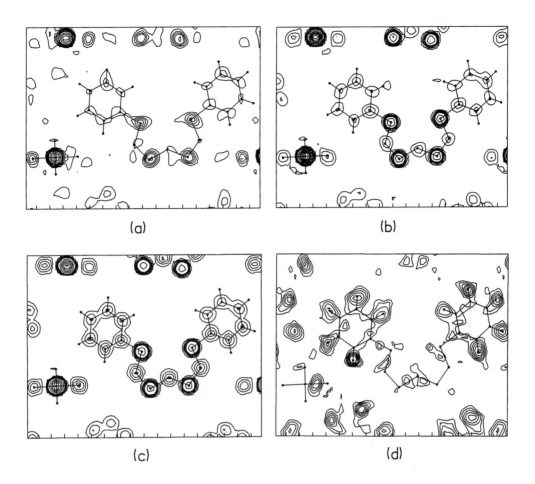

Fig. 2.19 Successive electron density and difference electron density syntheses in the development of the structure of [PhSNSNSNSPh]$^+$ [AsF$_6$]$^-$ starting from the position of the arsenic atom derived from a Patterson map. The contour interval is much smaller for the difference electon density map (d).

2.10   Refining the structure

Once all the non-hydrogen atoms have been found, the model structure needs to be *refined*. This means varying the numerical parameters describing the structure to produce the 'best' agreement between the diffraction pattern calculated from it by a Fourier transform and the observed diffraction pattern. Since there are no observed phases, the comparison of observed and calculated diffraction patterns is made entirely on their amplitudes $|F_o|$ and $|F_c|$. Changing any of the structural parameters (modifying the model structure in any way) affects the $|F_c|$ values, while the $|F_o|$ values remain fixed during the process.

The refinement process uses a well-established mathematical procedure called *least-squares* analysis, which defines the 'best fit' of two sets of data (here $|F_o|$ and $|F_c|$) to be that which minimizes one of the least-squares sums

$$\sum w(|F_o| - |F_c|)^2$$
$$\text{or } \sum w(F_o^2 - F_c^2)^2 \tag{2.10}$$

The first of these (refinement on F) has historically been most commonly used, but the second (refinement on F^2) is now increasing in popularity and is, in many ways, superior. The contribution of each reflection to the sum is weighted according to its perceived reliability, usually with weights based on the experimental s.u.'s, such as $w = 1/\sigma^2(F_o^2)$ for refinement on F^2.

The least-squares refinement of crystal structures is similar, in principle, to finding a 'best-fit' straight line through a set of points on a graph, but is more complicated because (i) there are many variable parameters instead of just two (the gradient and intercept) for a straight line graph and (ii) the equation relating data to parameters (the Fourier transform) is far from linear. Because of the non-linearity, an approximate solution (the model structure) must be known before refinement can begin, and each least-squares calculation is approximate, not exact, giving an improvement to the model, but not the best possible fit; the calculation must be repeated several times until eventually the changes in the parameters are insignificant.

What are the numerical parameters to be refined? They are, for the most part, the terms describing the positions and vibrations of the atoms in the Fourier transform equation (1.17). For each atom there are three positional coordinates x, y, z and a displacement parameter U, which can be interpreted as an *isotropic mean-square amplitude of vibration* (in Å^2) of the atom. In most experiments a significantly better fit to the data can be achieved by using more than one displacement parameter per atom in the model structure, allowing each atom to vibrate by different amounts in different directions (*anisotropic vibration*). The usual mathematical treatment has six U values (one for each axis and three cross-terms) for each atom in order to give different vibration amplitudes in three orthogonal directions which are, in general, not along the unit cell axes (Fig. 2.20). Thus, there are commonly nine refined parameters for each independent atom (atoms which are not related to each other by symmetry) in the structure. In addition, a scale factor has to be refined, which puts the $|F_o|$ and $|F_c|$ values on the same scale (the $|F_o|$ scale is arbitrary at the time of measurement, but $|F_c|$ values are calculated relative to the scattering power of one electron). There may be a small number of other refined correction factors, but these are not important for a basic understanding of the procedure.

Although there are many parameters to be refined for all but the smallest structures, the diffraction experiment usually provides an even greater number of observed data, unless X-ray scattering is unusually weak. Typically, the

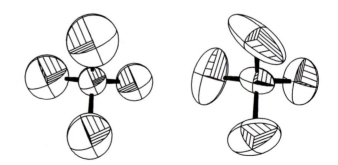

Fig. 2.20 Isotropic (left, represented as spheres) and anisotropic (right, represented as ellipsoids) atomic displacements for a perchlorate anion.

data/parameter ratio exceeds 6, and it may be as high as 20. The structure refinement problem is said to be 'over-determined', and this is essential in order to produce precise (reliable) parameters. As well as providing a value for each refined parameter, the least-squares process also gives a standard uncertainty. These parameter s.u.'s depend on the s.u.'s of the data (a good structure requires good data!), on the extent of agreement of the observed and calculated data (a lower least-squares sum gives lower parameter s.u.'s; another function closely related to this sum is called the *goodness of fit*), and on the excess of data over parameters (a greater excess gives lower parameter s.u.'s). Both the quality and the quantity of measured data matter for the quality (reliability) of the structure derived from them.

Once the model structure has been refined with anisotropic displacement parameters for the atoms, it is often possible to see small but significant difference electron density peaks in positions close to those expected for hydrogen atoms, particularly if there are few or no heavy atoms in the structure. Hydrogen atoms are more likely to be located from measurements taken at low temperature, because this reduces the vibration of the atoms and sharpens the electron density peaks. It is possible to include the hydrogen atoms in the refinement, and this may improve the fit slightly, but their parameters usually have large s.u.'s because their low electron density means they contribute only weakly to the diffraction of X-rays, so the measured intensities are relatively insensitive to the hydrogen atom parameters. In most cases, refinement is more successful if *constraints* are applied to hydrogen atom parameters, e.g. by keeping their bond lengths fixed and by tying their U values to those of the atoms to which they are attached. Details of how this is done vary enormously with different computer programs and with the habits and preferences of different crystallographers; some examples can be seen in the next chapter.

The refinement stage usually involves the vast majority of the computing resources used in a crystal structure determination, simply because the calculations are many and very repetitive. Compared with finding the atoms initially, it is generally a much less interesting process, but its correct execution is very important, since it delivers the final parameters describing the structure.

At the end of refinement, a difference electron density map should contain no significant features (peaks or holes). This calculation is usually performed as an extra check on the validity of the refined model structure. Typically, a final map with no features outside the range ± 1 e Å^{-3} is accepted as one evidence of a satisfactory structure determination.

2.11   Presenting and interpreting the results

What in fact are the results of a crystal structure determination? Returning to the microscope analogy, the application of the reverse Fourier transform equation to the observed diffraction pattern (but using calculated rather than genuinely observed phases!) gives an electron density map, an image of the X-ray scattering power of the crystal sample. It is, however, rare for the results to be presented in this way. Instead, the structure is represented as atoms (positioned at the centres of peaks of electron density) joined together by chemical bonds, and these atoms are described numerically by the refined parameters of the model structure.

As a rough rule of thumb, around 100 data per non-hydrogen atom in the asymmetric unit should be ample.

The *goodness of fit* is another standard statistical parameter, intended to show how well the calculated diffraction pattern corresponding to the model structure agrees with the observed diffraction pattern. For an ideal agreement and a correct weighting scheme, the goodness of fit should have a value of unity; considerable variation is observed in practice.

Constraints are conditions which are imposed on the refinement, for example by requiring certain parameters to have particular values rather than being free to take values which give the best agreement between observed and calculated diffraction patterns. Constraints may be imposed for various reasons, including the requirements of symmetry or the need to control parameters which are poorly defined by the diffraction data.

48   *The method step by step*

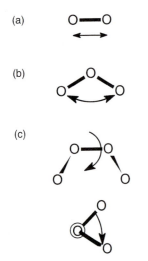

Fig. 2.21 Geometrical parameters:
(a) bond length; (b) bond angle;
(c) torsion angle (two views, the second
down the central bond).

The primary results from the refinement are the unit cell geometry and symmetry (space group), and the positions of all the atoms in the unit cell (three coordinates each), together with their isotropic (one) or anisotropic (six) displacement parameters. The displacement parameters are usually interpreted as representing thermal vibration of the atoms and, in most cases, are regarded as less important and less interesting than the positional parameters; they are also more affected than the positional parameters by many experimental errors.

From the atomic coordinates, unit cell geometry and symmetry, many geometrical results can be derived. These include

(a) *bond lengths* (distance between two atoms considered to be bonded together; see Fig. 2.21(a); a normal X-ray diffraction experiment does not directly show bonds, which are an interpretation based on distances and chemical experience);

(b) *bond angles* (angle between two bonds at one atom; see Fig. 2.21(b));

(c) *torsion angles* (the apparent angle between two bonds A–B and C–D when viewed along the B–C bond for a connected sequence of atoms A–B–C–D; see Fig. 2.21(c));

(d) the shapes and *conformations* of rings (e.g. chair and boat conformations of cyclohexane rings);

(e) the *planarity* or otherwise of groups of atoms (with possible consequences for the interpretation of their bonding);

(f) *degree of association* (monomers, formation of small oligomers, polymers);

(g) *intermolecular* geometry such as hydrogen bonding, van der Waals contacts, π-interaction stacking of planar aromatic groups.

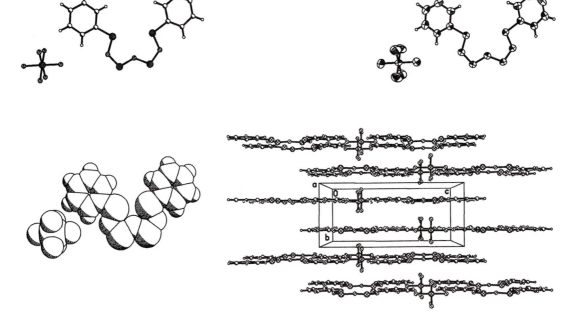

Fig. 2.22 Various styles of pictorial representation of the structure of $[PhSNSNSNSPh]^+[AsF_6]^-$: top left, conventional ball-and-stick model; top right, atomic displacement ellipsoids; bottom left, space-filling model; bottom right, packing of cations and anions in parallel layers in the crystal structure.

As well as numerically, the results may be displayed graphically, as pictures of individual molecules and of the packing arrangement of molecules in the crystal structure (Fig. 2.22). Since these are all interpretative models, not direct observations (unlike what is seen through a standard optical microscope), a wide variety of styles of representation is possible, the traditional *ball-and-stick* model being the most commonly used. It is also possible, of course, to construct accurately scaled three-dimensional models of structures from the atomic coordinates, though this is likely to be very time-consuming! In terms of the microscope analogy, the effective magnification for a typical molecule is $> 10^8$.

Further interpretation and explanation of the structure and its relation to physical and chemical properties then follows as appropriate. For a large and complex structure, this can be quite a task.

The amount and detail of structural information produced is greater than for any spectroscopic method of investigating chemical structure. It is salutary to recall that it all comes from a sample a fraction of a millimetre in size. Such is the power of the technique of crystal structure determination.

This raises other questions, for example whether the particular crystal selected is actually representative of the bulk sample, which may not be a pure homogeneous compound.

3 The method illustrated by examples

In this chapter the process of crystal structure determination is illustrated by a series of examples drawn from a wide range of structural chemistry research. The examples have been chosen to cover many different aspects of the experimental measurements and methods of structure solution and refinement, as well as a variety of types of material. Not all details of every structure determination are given, but each example presents particular features of interest. At the end of the chapter (section 3.8) some problems are presented for the reader to solve.

3.1 Case study 1: a mercury thiolate complex

The complex $[Et_4N][Hg(SR)_3]$, where R is the *cyclo*-hexyl group C_6H_{11}, is prepared from $HgCl_2$, NaSR and $[Et_4N]Cl$ in acetonitrile solution. Examination of a crystal of size $0.52 \times 0.36 \times 0.34$ mm on a four-circle diffractometer with molybdenum radiation of wavelength 0.71073 Å, at a temperature of 240 K, reveals a triclinic unit cell with dimensions

$$a = 10.724(4) \qquad b = 12.440(5) \qquad c = 12.643(5) \text{ Å}$$
$$\alpha = 72.40(2) \qquad \beta = 79.36(2) \qquad \gamma = 73.33(2)°$$
$$V = 1531.3(10) \text{ Å}^3$$

The formula mass for $C_{26}H_{53}HgNS_3$ is 676.5 daltons; this gives a calculated density of 1.467 g cm^{-3} and an average volume of 24.7 $Å^3$ per non-hydrogen atom, both of which are reasonable for such a compound, if $Z = 2$. This means there are two cations and two anions in the unit cell. There are only two possible triclinic space groups (see section 1.5), and the more likely is $P\bar{1}$, which requires the two cations to be related to each other by inversion symmetry, and similarly for the two anions in the unit cell; the asymmetric unit of the structure is one cation and one anion, so we know nothing at this stage about the molecular geometry from symmetry arguments.

All possible reflections with $\theta < 25°$ have been measured, a few of them more than once, giving a total of 10 990 reflections. Corrections are applied for absorption effects, which are strong for a compound containing mercury, based on measurements of intensities of selected medium–strong reflections at a range of different crystal orientations (these would be equal if there were no absorption and a correction can be calculated from their observed variation); it is also found that the intensities have decreased steadily by about 7% during the data collection period of around 4 days, and this is corrected for. Each pair of reflections with indices h, k, l and $-h, -k, -l$ is equivalent by symmetry, so they are averaged, to give a unique set of 5412 reflections. The averaging process provides a measure of the agreement of symmetry-equivalent data in

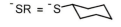

Fig. 3.1 The cyclohexanethiolate ligand.

The numbers in parentheses are standard uncertainties, expressed for compactness as units in the last figure of the corresponding numerical value. Thus, for example, 10.724(4) Å means a value of 10.724 Å with a s.u. of 0.004 Å.

In this space group symbol, the *P* means a primitive unit cell, and the $\bar{1}$ means an inversion centre as the only other symmetry.

the form of a factor R_{int}, defined rather like the R factor in structure refinement, except that comparison is between pairs of observed symmetry-equivalent reflections instead of between observed and calculated values; the value 0.022 obtained for this set of data is excellent.

With just one heavy atom in the asymmetric unit, two in the unit cell, this structure is an obvious candidate for a Patterson synthesis as the means of solution. The largest peaks found in the Patterson synthesis (one half of the unit cell only; the other half is equivalent by inversion symmetry) are shown in Table 3.1. The peak heights are scaled arbitrarily so that the largest peak, at the origin, has a height of 999. The length given for each peak is the length of the corresponding interatomic vector, which is the distance (in Å) between the two atoms concerned.

The subscript 'int' stands for *internal*, since this is a measure of the internal consistency of agreement of the data, not agreement with something else.

Table 3.1 The largest Patterson peaks for case study 1

Peak number	x	y	z	Peak height	Vector length (Å)
1	0.000	0.000	0.000	999	0.00
2	0.462	0.146	0.432	403	8.87
3	0.354	0.273	0.265	132	7.49
4	0.111	0.867	0.163	118	2.47
5	0.466	0.957	0.407	115	7.61
6	0.003	0.811	0.975	110	2.46
7	0.074	0.022	0.158	101	2.39
8	0.462	0.836	0.411	99	7.28

In this space group, for each atom at a position x, y, z there is a symmetry-equivalent atom at position $-x$, $-y$, $-z$. The two mercury atoms in the unit cell thus have coordinates which are equal and opposite in sign, and the vector between these two positions is $x - (-x)$, $y - (-y)$, $z - (-z)$, which is just $2x$, $2y$, $2z$. The highest Patterson peak, therefore, excluding the origin peak, should correspond to a Hg—Hg vector, should be much larger than the other peaks, and has coordinates equal to twice those of a mercury atom; its vector length is the distance between the two mercury atoms in the unit cell. The coordinates of one mercury atom are thus 0.462/2, 0.146/2, 0.432/2, giving $x = 0.231$, $y = 0.073$, $z = 0.216$. The two mercury atoms are well separated (almost 9 Å apart); a short distance here would indicate that the two mercury atoms are in fact part of a dimeric anion of formula $[Hg_2(SR)_6]^{2-}$, probably with bridging thiolate ligands.

A mercury atom has 80 electrons, and this is a significant proportion of the total scattering power of the asymmetric unit (344 electrons). Since we now know the position of the mercury atom, we can use this as our first model structure in the next stage of Fourier syntheses to find the remaining atoms. It is, however, worth pausing to examine the next highest peaks in the Patterson map. There are six of these with similar heights; all other peak heights are less than 60 on this scale. The next largest peaks after Hg—Hg are expected to be Hg—S. There should be three of these corresponding to *intramolecular* vectors, i.e. Hg—S bonds within one anion, and three corresponding to *intermolecular* vectors, i.e. from mercury in one anion to the three sulfur

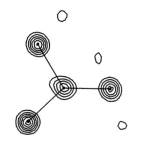

Fig. 3.2 Part of the difference electron density map around the mercury atom position; the points and lines show the final refined mercury and sulfur atom positions and bonds.

atoms in the other anion. The list does include three vectors of length 2.47, 2.46 and 2.39 Å, which are appropriate for bonds, and three vectors more than 7 Å long, which will be the intermolecular ones. So the Patterson map is certainly consistent with our proposed chemical formula and with a three-coordinate Hg-centred anion.

Now we move into the structure completion 'bootstrap' procedure. Taking the mercury atom alone as the model structure, Fourier transformation gives a calculated diffraction pattern. The value of the residual factor R is 0.2840 for the 5032 reflections which have $F^2 > 2\sigma(F^2)$ (such reflections are sometimes called '*observed reflections*', because they have an intensity judged to be significantly higher than background scattering), and $wR2$ is 0.6497 for the complete set of reflections. A difference electron density map calculated from the observed amplitudes and the calculated phases derived from just the mercury atom does not show the full electron density of the mercury atom (it would appear clearly as a very large peak in a full electron density map, but a difference map is better for finding new atoms). Its highest three peaks, with electron densities above 20 e Å^{-3}, are in positions about 2.4 Å from mercury, suitable for sulfur atoms; all 26 carbon atoms and the single nitrogen atom are among the 32 next highest peaks excluding the residual peak at the position of mercury, with densities in the range 8.6 down to 2.8 e Å^{-3}, and the remaining peaks are rather lower. It is not expected that all carbon atoms will have the same electron density, because those which undergo larger vibrations have their electron density spread out over a larger volume, so they usually show up as smaller peaks. The assignment of atom types is made on the basis of the observed distances and angles involving the peak positions and the expected structure, as well as on peak heights.

So in this case, a single cycle of Fourier synthesis calculations reveals all the non-hydrogen atoms, and refinement of the structure can begin. Refinement with isotropic displacement parameters for all the atoms (each atom has one adjustable overall displacement parameter as well as three adjustable coordinates) reduces $wR2$ from 0.3647 (with all atoms in the positions found from the difference map) to 0.3112 and the value of R after refinement is 0.1021, a considerable improvement on the first trial structure with just the mercury atom present. Inclusion of anisotropic displacement parameters (six values for each atom instead of one) reduces $wR2$ to 0.1371 and R to 0.0406; there are now 280 refined parameters.

Most of the 53 hydrogen atoms now show up in a further difference electron density map, those in the anion being clearer than those in the cation, for which the atoms have higher displacement parameters, so their electron density is more spread out. They are included in the refinement, but the C—H bond lengths and angles involving hydrogen atoms are kept fixed (constrained) at ideal values rather than being allowed to refine freely, because the hydrogen atoms are not very precisely located by X-ray diffraction; effectively, the hydrogen atoms are made to ride on their parent carbon atoms. This technique incorporates all the electron density of the atoms in the model structure, but it adds very few or no extra refined parameters; in this particular structure, free rotation is allowed about the C—C bonds of the cation starting from an idealized staggered conformation; such small deviations from ideal positions are due to intermolecular interactions. The final refinement also includes some extra minor corrections for effects which

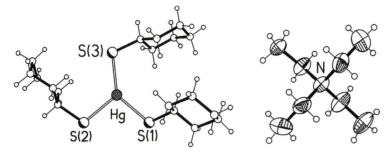

Fig. 3.3 The structure of the cation and anion of case study 1; the anion is shown in ball-and-stick form, the cation with displacement ellipsoids for carbon and nitrogen atoms, to illustrate different styles of representation.

are not important for this account, including small modifications to the relative weighting of different reflections, and gives values of 0.0823 for *wR2* (all data) and 0.0315 for *R* (observed data). There are 285 refined parameters derived from the 5412 data, a very high degree of over-determination, so the precision of the structure is high (s.u. values of the parameters are small). In a final difference electron density map there a few peaks of size 0.8–2.1 e Å^{-3} very close to the mercury atom (a commonly observed feature for heavy atom structures, due largely to an imperfect correction for strong absorption effects in the data), and no other peaks above 0.6 e Å^{-3}, which is an insignificant level.

The cation and anion are shown in Fig. 3.3. The main interest is in the coordination of the mercury atom, which is somewhat distorted trigonal planar, with a range of about 0.08 Å for the three Hg—S bond lengths, much greater than their individual uncertainties of about 0.002 Å, and three very different S—Hg—S angles, the smallest being 101.27(5)° and the largest 135.82(5)°. As expected, all the cyclohexane rings show a chair conformation with sulfur in an equatorial position. There is nothing at all unexpected or special about the geometry of the cation, which occurs in many salts of complex anions. It is an important feature of crystal structure determination that, in general, the whole structure has to be determined, even if only one particular part of it is really of interest; it is an all-or-nothing technique. This is a direct consequence of the nature of Fourier transforms: all the atoms in the unit cell contribute together to the observed diffraction pattern, and all the diffraction pattern has to be used to find the atoms.

In summary, this case study illustrates particularly how simple it is to find a single heavy atom in the asymmetric unit of this common low-symmetry space group by a Patterson synthesis, the straightforward completion of the structure from this one heavy atom as a starting point, and the necessity to locate and refine all the atoms of the structure even if some parts of it are of little or no particular interest.

3.2 Case study 2: a lithium complex as its iodide salt

This compound is the product of a rather complex reaction in which the reagents included lithium iodide (LiI), the tridentate nitrogen-donor ligand pmdeta (an abbreviation for its full chemical name), dibenzylamine (Bz₂NH) and *n*-butylsodium (a deprotonating agent). The composition of the product

Fig. 3.4 The molecules pmdeta (top) and dibenzylamine (bottom).

was not at all clear from chemical analysis and spectroscopic measurements, and suggestions included an empirical formula of $(Bz_2N)Li_2I(pmdeta)_2$.

Data were collected on an area-detector diffractometer with X-rays of wavelength 0.71073 Å and a crystal of size $0.43 \times 0.38 \times 0.38$ mm, at a temperature of 160 K. With this much faster instrument, a total of 8734 reflections were obtained overnight, giving 3424 unique data with $\theta < 25.6°$ and an internal agreement index $R_{int} = 0.036$. The expected presence of iodine in the structure means that absorption effects are significant, so a correction was applied as for the previous example. No intensity decay was observed in the short data collection time.

The crystal system is monoclinic and the space group is $P2_1/n$ (unambigously indicated by the systematic absences in the data; see section 1.5 for further information on this very common space group). The unit cell parameters are as follows:

$$a = 8.8507(9) \qquad b = 14.182(2) \qquad c = 16.371(2) \text{ Å}$$
$$\beta = 94.030(2)° \qquad [\alpha = \gamma = 90°] \qquad V = 2049.8(4) \text{ Å}^3$$

This cell volume is sufficient for two units of the suggested formula, but this is incompatible with the symmetry of the space group, which requires all atoms and groups of atoms to occur in sets of four unless they lie on inversion centres; this is not possible for a dibenzylamido group (Bz_2N), which is highly unlikely to be linear at the nitrogen atom. In this case, then, we must proceed without knowing the actual chemical contents of the asymmetric unit, which is one-quarter of the unit cell.

The presence of iodine (53 electrons; no other atom has more than seven) means a Patterson synthesis is again appropriate for the solution of this structure. For this space group there are four equivalent general positions in the unit cell, with coordinates x, y, z; $-x, -y, -z$; $\frac{1}{2} - x, \frac{1}{2} + y, \frac{1}{2} - z$; and $\frac{1}{2} + x, \frac{1}{2} - y, \frac{1}{2} + z$. Four atom positions give 16 vectors as shown in Table 3.2, where each entry in the table body is the difference between the position at the top of the column and the position at the left of the row; wherever the number $-\frac{1}{2}$ would appear by this simple subtraction, it is replaced by $\frac{1}{2}$, because it is always permissible to add or subtract any whole number (in this case adding one) to a coordinate, which has the effect of moving to an exactly equivalent position in another unit cell.

It is also possible to calculate the Patterson function in such a way that the origin peak is removed.

Each row, and each column, contains one entry 0, 0, 0; one entry with $\frac{1}{2}$ as the first and third coordinates; one entry with $\frac{1}{2}$ as the second coordinate; and

Table 3.2   Vectors for equivalent atoms in space group $P2_1/n$

	x,y,z	$-x,-y,-z$	$\frac{1}{2}-x,\frac{1}{2}+y,\frac{1}{2}-z$	$\frac{1}{2}+x,\frac{1}{2}-y,\frac{1}{2}+z$
x,y,z	$0,0,0$	$-2x,-2y,-2z$	$\frac{1}{2}-2x,\frac{1}{2},\frac{1}{2}-2z$	$\frac{1}{2},\frac{1}{2}-2y,\frac{1}{2}$
$-x,-y,-z$	$2x,2y,2z$	$0,0,0$	$\frac{1}{2},\frac{1}{2}+2y,\frac{1}{2}$	$\frac{1}{2}+2x,\frac{1}{2},\frac{1}{2}+2z$
$\frac{1}{2}-x,\frac{1}{2}+y,\frac{1}{2}-z$	$\frac{1}{2}+2x,\frac{1}{2},\frac{1}{2}+2z$	$\frac{1}{2},\frac{1}{2}-2y,\frac{1}{2}$	$0,0,0$	$2x,-2y,2z$
$\frac{1}{2}+x,\frac{1}{2}-y,\frac{1}{2}+z$	$\frac{1}{2},\frac{1}{2}+2y,\frac{1}{2}$	$\frac{1}{2}-2x,\frac{1}{2},\frac{1}{2}-2z$	$-2x,2y,-2z$	$0,0,0$

one entry with general coordinates. Some of the entries are identical, and others are equivalent to each other by symmetry: the 16 vectors from four symmetry-equivalent atoms actually give only nine different peaks (those which coincide produce correspondingly greater peak heights) and only four which are not directly related to each other by symmetry, these four peaks being the entries in any one row or any one column. So, if there is indeed one iodine atom in the asymmetric unit of the structure, we will expect to find in the Patterson synthesis a very large origin peak (as always); a large peak with x and z coordinates both equal to $\frac{1}{2}$ and a y coordinate from which we can easily calculate the y coordinate of the iodine atom; another large peak with $\frac{1}{2}$ for its y coordinate and with x and z coordinates from which we can similarly find the x and z coordinates of the iodine atom; and a peak of about half the size of these two with coordinates equal to twice the x, y, z coordinates of the iodine atom. Other peaks will be significantly lower.

The highest peaks in the Patterson synthesis calculated from the diffraction pattern of this compound are listed in Table 3.3. All other peaks are under 80 in height. Because of the monoclinic symmetry, there are other peaks equivalent to those listed in the table; for example, there are three other peaks equivalent to peak number 4.

Table 3.3 The largest Patterson peaks for case study 2

Peak number	x	y	z	Peak height	Vector length (Å)
1	0.000	0.000	0.000	999	0.00
2	0.500	0.208	0.500	443	9.50
3	0.014	0.500	0.416	422	9.83
4	0.486	0.292	0.084	213	6.06

Finding the position of the iodine atom in the asymmetric unit from these peaks is a matter of identifying each of the peaks with a corresponding entry in Table 3.2. There are in fact several possible solutions, all of which are entirely equivalent and equally correct. Thus, for example, peak number 2 can be identified with the entry $\frac{1}{2}, \frac{1}{2} - 2y, \frac{1}{2}$ and this means $\frac{1}{2} - 2y = 0.208$, so $y = 0.146$. Peak number 3 can be identified with the entry $\frac{1}{2} - 2x, \frac{1}{2}, \frac{1}{2} - 2z$, which gives $x = 0.243$ and $z = 0.042$. These three coordinates for the iodine atom agree with peak number 4, identified with the entry $2x, 2y, 2z$. The shortest distance between two iodine atoms in the structure is the shortest vector length for the three peaks, which is over 6 Å.

The iodine atom serves as our first model structure, and the process of structure completion by Fourier calculations and examination of electron density syntheses (or difference electron density syntheses) proceeds as for the first case study. With just the iodine atom, $wR2$ and R are 0.6904 and 0.2850 respectively. The first 21 peaks in a difference map have heights ranging from 8.83 to 4.04 e Å^{-3}, and the remaining peaks are under 2.6 e Å^{-3}. These 21 peaks can all be assigned atom types based on the observed geometry, which reveals a pmdeta molecule, a lithium atom and a BzN fragment (nitrogen with just one benzyl group attached, not two).

Inclusion of all these atoms in a new model structure with isotropic displacement parameters for all atoms reduces $wR2$ to 0.2523 and R to 0.0683; refinement with anisotropic displacement parameters gives further reduction to 0.1938 and 0.0463. At this point a difference electron density map shows 32 hydrogen atoms in sensible positions: peak heights for these are between 0.75 and 0.50 e Å^{-3}, and the next highest peak is 0.44 e Å^{-3}. There are two hydrogen atoms attached to the nitrogen atom of the BzN group (which is coordinated to lithium), so this is shown to be a benzylamine ligand, BzNH$_2$. In the subsequent refinement, the positions of these two hydrogen atoms are left to refine freely, in order not to impose a preconceived idea of the geometry at the nitrogen atom, but all the hydrogen atoms bonded to carbon atoms are constrained to the expected geometry.

The total number of refined parameters is 211, so we still have a very high data/parameter ratio. The final values of $wR2$ and R are 0.0595 and 0.0255, and a difference electron density map shows no peaks higher than 0.32 and no holes deeper than 0.37 e Å^{-3}, indicating that the atoms are all correctly accounted for.

The asymmetric unit of the structure is shown in Fig. 3.5. The compound is the iodide salt of a lithium-centred complex cation having as ligands a tridentate neutral pmdeta molecule and a benzylamine molecule. The reaction to produce it must involve cleavage of a C—N bond of the original dibenzylamine reagent. Lithium has an approximately tetrahedral coordination geometry, which is not unusual.

Examination of the relative positions of the cation and anion, and symmetry equivalents of them, shows that the hydrogen atoms of the NH$_2$ amine group are both directed towards iodide anions to give N—H...I hydrogen bonding. Each iodide anion is involved in two such hydrogen bonds, to give dimeric units of two cations and two anions, as shown in Fig. 3.6. These dimeric units are packed together in the crystal structure as shown in Fig. 3.7, with normal van der Waals interactions holding the units together.

In summary, this example illustrates a typical Patterson solution for a structure with one heavy atom in the asymmetric unit, free refinement of some hydrogen atoms positions from high-quality data, the occurrence of hydrogen

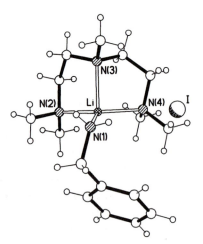

Fig. 3.5   The asymmetric unit in the structure of case study 2.

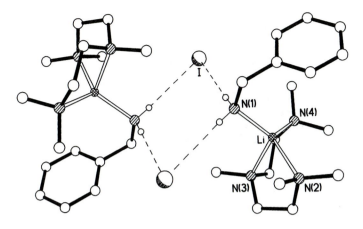

Fig. 3.6 Dimeric units (two cations and two anions) held together by hydrogen bonding in the structure of case study 2.

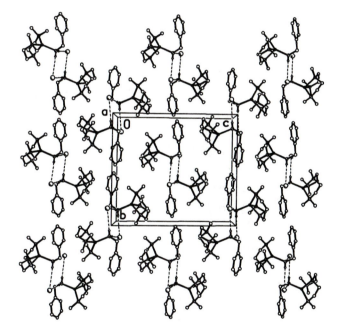

Fig. 3.7 The packing of dimeric units in the crystal structure of case study 2.

bonding in a structure, and the successful application of crystal structure determination to a compound of hitherto unknown composition.

3.3 Case study 3: another closely related lithium complex

The unexpected result of the previous investigation prompted further study of related systems, and it was found that high yields of the same material could be obtained by simple reaction of LiI, pmdeta and benzylamine (BzNH$_2$) in toluene solution. Use of *p*-methylbenzylamine (Fig. 3.8) instead of benzylamine gave a product for which chemical analysis and spectroscopic measurements were consistent with an analogous formula [(pmdeta)-

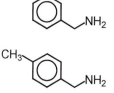

Fig. 3.8 benzylamine (top) and *p*-methylbenzylamine (bottom).

Li(H$_2$NCH$_2$C$_6$H$_4$Me)]I. This, together with other substituted versions, was structurally characterized in order to explore the effects of different substituents on the amine ligand.

Crystallographic measurements were carried out as for the previous example, with a crystal of size 0.64 × 0.47 × 0.16 mm and a temperature of 160 K. The unit cell is orthorhombic and body-centred with two-fold rotation symmetry along each of the three unit cell axes but no reflection or inversion symmetry: the space group is *I*222. There are in fact no fewer than six different space groups all theoretically possible from inspection of the diffraction data (six different possible arrangements of rotation/reflection/inversion symmetry elements, all consistent with the observed diffraction pattern), and the correct one must be chosen by trying each one in turn and seeing which gives a recognizable solution. In fact, the correct space group is judged to be the most likely one on the basis of some statistical criteria, and was the first one attempted; this was successful, so it was not necessary to consider the others further. Cell parameters are as follows.

$$a = 13.5819(11) \qquad b = 13.8836(11) \qquad c = 23.593(2) \text{ Å}$$

$$[\alpha = \beta = \gamma = 90°] \qquad V = 4448.8(6) \text{ Å}^3$$

This unit cell volume is sufficient for eight formula units, with an average volume of 24.2 Å^3 per non-hydrogen atom (rather higher than the notional 18, but typical for this area of structural chemistry) and a calculated density of 1.278 g cm^{-3}.

As for the previous example, a correction was applied for absorption effects, but was unnecessary for intensity decay. A total of 14 059 measured reflections with $\theta < 28.4°$ gave 5050 unique data after averaging of symmetry equivalents, with a value of 0.025 for R_{int}.

In this case, for variety, the structure solution is carried out by direct methods. In a fully automatic mode, the computer program selected 252 relatively intense reflections with an impressive 15 847 relationships among their phases. Sixty-four different sets of starting phases were generated at random and were adjusted to give the best agreement with these relationships in each case; 14 of them converged to essentially the same set of well-fitting phases. Note that, for a structure with no inversion symmetry, each phase can take any value from 0 to 360°, and these calculations are more time-consuming than they would be for a centrosymmetric structure. An electron density map calculated from the observed amplitudes and estimated phases of the strongest reflections shows two large peaks, both lying on two-fold rotation axes (which have x and y coordinates of 0 and $\frac{1}{2}$), and no other geometrically recognizable fragments. An iodine atom lying on a rotation axis of symmetry has only three other symmetry equivalents in the unit cell, whereas an atom in a general position has seven others, because the rotation symmetry relates the atom to itself and not to a different one. Two iodine atoms on rotation axes thus give $Z = 8$ as expected.

Alternatively, exactly the same solution for the iodine atoms can be found from a Patterson synthesis, which shows four large peaks with heights between 434 and 181 (relative to the origin peak of height 999), from which the coordinates of the atoms, related to each other by the two-fold rotation axes, are easily deduced.

In the space group symbol *I*222, the letter *I* (for the German word *innenzentriert*) means a body-centred unit cell, and each digit 2 stands for a two-fold rotation axis; for any orthorhombic space group, the symbol gives the unit cell centring type, followed by the symmetry corresponding to each of the *a*, *b* and *c* axes.

This means we are trying to find 252 unknown values, and we have 15 847 pieces of information about combinations of these values. Each of these pieces of information is approximate and has an associated probability of being true (a reliability). The ratio of information (relationships) to unknowns (phases) is very high, which it needs to be for any chance of success.

Use of the two iodine atoms as the initial model structure leads to a problem. Because these atoms lie in special positions on symmetry elements, their arrangement in the unit cell, together with their equivalents, actually has higher symmetry than the complete structure with all atoms present; the apparent space group is a different one, in which there are also mirror planes. This incorrect extra symmetry carries through the Fourier transform equations into the calculated diffraction pattern and hence, via the calculated phases (which are all either 0 or 180° because of the extra symmetry, when they should really be able to take any value between 0 and 360°), back into the next electron density map. The values of $wR2$ and R are 0.5976 and 0.1987. In the electron density map two amine ligands are seen, related to each other by mirror symmetry which does not belong to the space group (Fig. 3.9), and two pmdeta molecules are also visible in the same way; the relatively electron-poor lithium does not show up clearly at this stage. Only one of the amine ligands can be correct, and only one of the pmdeta molecules, and there is a 50% probability of choosing a valid combination. In such cases, it is simplest to select only some of the possible atoms, keeping to those which we are sure belong together. For this structure one of the amine ligands is chosen (they are equally acceptable), giving us one nitrogen and eight carbon atoms to add to the model structure.

This model structure no longer has false mirror planes. The calculated amplitudes are closer to the observed ones ($wR2 = 0.5181$, $R = 0.1642$), and the calculated phases are more nearly correct, so the next electron density map is clearer, showing the pmdeta molecule and the lithium atom, with only small peaks corresponding to mirror images of these and of the amine ligand. For this structure, two cycles of the 'bootstrap' Fourier procedure have been required in order to find all the non-hydrogen atoms.

Refinement now proceeds as before, with reduction of $wR2$ and R to 0.3251 and 0.0862 respectively with isotropic displacement parameters, and then to 0.1618 and 0.0406 with anisotropic parameters. Most of the hydrogen atoms show up in a difference electron density map. They are included in the refinement in the same way as for the previous example. The difference map also shows that two of the CH_2 groups of the pmdeta ligand do not actually lie in the same position in every molecule, but that each of them occupies two alternative positions at random; that is, for each of them, one position is adopted in some of the molecules, and the other position is adopted in the rest. This is by no means an uncommon observation, and is referred to as disorder in the structure. It means that all the molecules are not actually identical, but there is some variation (in this case a choice of two alternatives for each of two groups of atoms), and X-ray diffraction gives us only an average structure, which appears as partial atoms. This *disorder* is included in the refinement, with the distribution ratio between the two alternative sites as a variable parameter. The final values of $wR2$ and R are 0.0543 and 0.0221 respectively. The total number of refined parameters is 241, and there are no features in a final difference map outside the range ± 0.54 e Å^{-3}.

There is one other important feature of this structure determination. We have previously seen that a diffraction pattern always has a centre of symmetry, even if the crystal structure is not centrosymmetric; this is known as *Friedel's law*. In fact, this is only approximately true, because of an effect known as *anomalous scattering*. As a first approximation, every time an atom

Fig. 3.9 A benzylamine group and its false mirror image, both of which appear together in the first difference map for case study 3.

There *are* automatic computer programs to solve this problem.

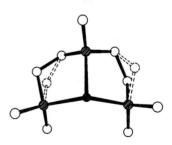

Fig. 3.10 Disorder in the ethylene linkages of pmdeta in case study 3; the minor component of each linkage is shown dotted.

scatters X-rays, a phase shift of 180° occurs; because this phase shift is constant, it can be ignored, and we regard all atoms as scattering in phase at $\theta = 0°$. In reality, the phase shift is not exactly 180°, and it is different for different atoms, generally increasing with atomic number, although there are irregularities in the pattern; atoms which contribute strongly to X-ray absorption also give significant anomalous scattering. For centrosymmetric structures, the effects of anomalous scattering on the pair of opposite reflections h, k, l and $-h, -k, -l$ are equal, and so they still do have the same intensity: Friedel's law is obeyed. For non-centrosymmetric structures, however, the effects do not cancel, and these reflections, known as *Friedel pairs* or *Friedel opposites*, have different intensities. The differences are not very great in most cases, since anomalous scattering is only a small fraction of the total atomic scattering of X-rays, but careful measurement and comparison of Friedel pairs of reflections, or inclusion of them as separate non-equivalent data in refinement, allows us to distinguish a crystal structure from its inverse or opposite hand. For *chiral* molecules, this represents a direct experimental method of determining *absolute configurations*, which is not possible otherwise.

A *chiral* molecule is one which is not identical to its mirror image; the two non-identical mirror images, known as *enantiomers*, are related like left and right hands. Determining the *absolute configuration* means finding out exactly which one of the two we actually have.

Table 3.4   Normal and anomalous scattering factors

Element	Normal $f(0)$	$\Delta f''$ (Cu)	$\Delta f''$ (Mo)
Carbon	6	0.009	0.002
Nitrogen	7	0.018	0.003
Iodine	53	6.835	1.812

Table 3.4 gives the normal scattering power at zero Bragg angle (equal to the number of electrons) and, on the same scale, the amount of scattering 90° out of phase for carbon, nitrogen and iodine atoms irradiated by the most commonly used types of X-rays, derived from copper and molybdenum X-ray tube targets. It can be seen that, in the present structure, anomalous scattering effects are significant, and the correct *absolute structure* (a general term which includes absolute configuration and related expressions appropriate for structures which are non-centrosymmetric but not chiral) is indicated with confidence. It can be assessed (among other possible methods) by refinement of a special parameter which should be 0 for the correct result and 1 for the wrong absolute structure; for this structure, the parameter is –0.01(2), i.e. it is less than 1 s.u. from zero and so is definitely correct. Alternatively, both the structure and its inverse can be separately refined and the results compared; for this structure, refinement with everything inverted gives significantly higher values of $wR2$ (0.0697 compared with 0.0543) and R (0.0289 compared with 0.0221).

The asymmetric unit (consisting of a complex cation in a general position, and two iodide anions in special positions, each effectively serving as half an anion in this asymmetric unit and half in another) is shown in Fig. 3.11. It looks very similar to the previous structure (case study 2). However, the hydrogen bonding pattern is different. Instead of dimeric units, infinite chains are formed, each iodide linking to two cations and each amine NH_2 group linking to two anions, as shown in Fig. 3.12.

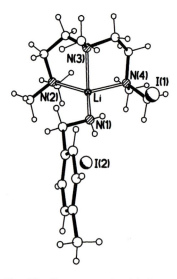

Fig. 3.11   The asymmetric unit in the structure of case study 3.

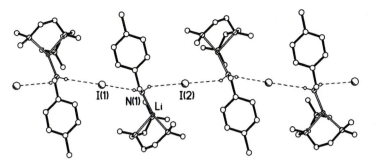

Fig. 3.12   Hydrogen-bonded chains of cations and anions in the crystal structure of case study 3.

In summary, this example illustrates a partial solution derived by direct methods, a *pseudosymmetry* problem caused by heavy atoms which in themselves have more symmetry than the complete structure, the occurrence of a small amount of disorder in a structure, and the use of anomalous scattering effects to determine absolute configuration (absolute structure). In conjunction with the previous example, it also illustrates how compounds which are chemically extremely similar can have significant differences in their crystal structures.

Indeed, it is also possible for one and the same compound to adopt more than one crystal structure under different conditions of crystallization (such as different rates of crystallization, different solvents, different temperatures, etc.). This phenomenon is called *polymorphism*, and is by no means uncommon. At an elementary level it is familiar in the different crystalline forms of sulfur (monoclinic and orthorhombic). It is common in minerals (such as calcite and aragonite, two crystalline forms of calcium carbonate), and it is commercially important in pharmaceuticals, where one polymorph of a substance may be an active drug while another polymorph of the same chemical substance is inactive or even harmful.

Pseudosymmetry means approximate but not exact symmetry; in this particular case, the arrangement of the heavy atoms alone fits a higher symmetry space group than that of the whole structure with all the light atoms included.

By its derivation, the word *polymorphism* means the existence of *many forms*; it is generally used when a compound is found in at least two different crystalline forms.

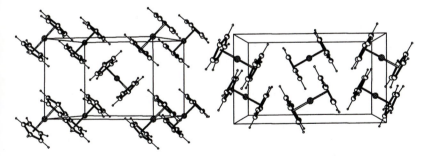

Fig. 3.13   Two polymorphic forms of ferrocene; they have different packing arrangements and different crystal symmetries, but the individual molecules are essentially the same in both structures.

3.4   Case study 4: a lithium borohydride complex

This material was prepared from $LiBH_4$ and pmdeta in toluene solution and a crystal of size $0.30 \times 0.24 \times 0.10$ mm was examined on a four-circle diffractometer, with copper radiation of wavelength 1.54184 Å (this sample gives relatively weak diffraction effects and an X-ray tube with a copper anode target gives higher intensity output), at a temperature of 200 K. The orthorhombic unit cell parameters are

$$a = 14.713(3) \qquad b = 11.820(2) \qquad c = 16.898(3) \text{ Å}$$

$$[\alpha = \beta = \gamma = 90°] \qquad V = 2938.7(9) \text{ Å}^3$$

and the space group is determined unambiguously from systematic absences as *Pbca*. The unit cell volume is sufficient for eight formula units, which is the number of general positions for this space group, so this tells us nothing about the molecular geometry.

Intensities have been measured to a maximum θ of 55°. Bragg angles are larger for copper than for molybdenum X-rays (see the Bragg equation: $\sin\theta$ is proportional to λ), and this corresponds to an angle of about 22° for molybdenum radiation; the intensities are much lower than for the previous case studies, and are too low to be significant at higher angles. No corrections are required for absorption or for intensity decay. The data set contains many symmetry-equivalent reflections; the 9583 measurements yield 1843 unique reflections after averaging of equivalents, with $R_{int} = 0.120$. One cause of this high value is the large number of weak (and hence not very precisely measured) reflections, such that only 837 reflections have $F^2 > 2\sigma(F^2)$.

There are no elements heavier than nitrogen in this compound, so we turn to direct methods for a solution. The 277 most significant reflections (highest intensities relative to the average for their particular Bragg angles) give 5757 phase relationships. Of 64 automatically generated starting phase sets, adjusted to give the best agreement of the phase relationships in each case, no fewer than 40 produce essentially the same electron density map. In this map there are three peaks with relative heights 260, 258, and 210, which are clearly nitrogen atoms when the geometry is examined. The boron atom corresponds to a peak with height 180. Carbon atoms show a higher degree of vibration and give peaks with heights 127–159, and the electron-poor lithium atom is the smallest peak for a genuine atom, at 114. All other peaks are below 62 in height, so the molecule stands out clearly from the background of the map (Fig. 3.14) and all non-hydrogen atoms are revealed in a single calculation.

Isotropic refinement reduces $wR2$ from 0.6063 to 0.4455, with $R = 0.1655$; addition of anisotropic displacement parameters gives $wR2 = 0.3397$, $R = 0.1260$. All the hydrogen atoms are now found in a difference map, with electron densities up to 0.35 e Å^{-3}. Those bonded to carbon are refined with riding-model constraints, but the borohydride hydrogen atoms are refined completely freely, since the geometry of this group is of primary importance in the study and should not be imposed with preconceived ideas! With optimization of the weighting, the final $wR2$ and R are 0.2362 and 0.0741 respectively, and there are no significant features remaining in a final difference map. A total of 149 parameters are refined.

The molecular structure is shown in Fig. 3.15. As in the previous examples, pmdeta functions as a tridentate ligand to lithium. The borohydride anion has a tetrahedral arrangement of hydrogen atoms around boron, as expected, and it is bonded to lithium through two bridging hydrogen atoms. The hydrogen atoms are less precisely located than the other atoms, of course, but the bonds between boron and the bridging atoms are slightly longer than those to the terminal hydrogen atoms.

This is another orthorhombic space group, so the symbol gives the lattice centring type (*P* for primitive), followed by the symmetry elements associated with each of the three unit cell axis directions in turn: a glide plane perpendicular to the *a* axis with translation in the *b* direction, one perpendicular to the *b* axis with translation in the *c* direction, and one perpendicular to the *c* axis with translation in the *a* direction.

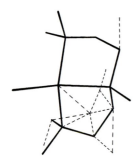

Fig. 3.14 Interpretation of the largest peaks in the map produced by direct methods for case study 4. Solid lines connect the highest 14 peaks, with dashed lines to the lower peaks.

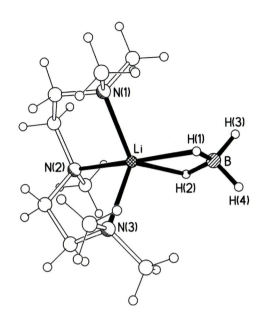

Fig. 3.15 The molecular structure for case study 4.

Figure 3.16 shows the displacement parameters of the structure, excluding hydrogen atoms bonded to carbon. Even at 200 K the vibration is considerable, which contributes to the relatively low diffraction intensities, especially at higher Bragg angles. Cooling below 200 K was not possible, as it led to a deterioration in the quality of the crystals. From data collected at room temperature it was not possible to locate the hydrogen atoms at all.

In summary, this example illustrates the significant scattering contribution that hydrogen atoms can make for light-atom structures, the effects of relatively high atomic vibration on diffraction intensities, the usefulness of copper radiation and low temperature data collection in such cases, and the use of direct methods for light-atom structures.

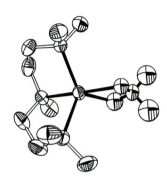

Fig. 3.16 Atomic displacement ellipsoids (spheres for the isotropic hydrogen atoms) for case study 4.

3.5   Case study 5: an indium complex

We have already looked at this example, in section 1.5, when considering preliminary unit cell and symmetry information. The proposed chemical formula is [(18-crown-6)K][In(NCS)$_4$(py)$_2$], where py is pyridine, and unit cell determination shows that there cannot be more than one formula unit in the cell. If the correct space group is the centrosymmetric $P\bar{1}$, which is much more common than the alternative non-centrosymmetric triclinic space group $P1$, then we already know that both the cation and the anion must lie on inversion centres, and that the arrangement of pyridine ligands around the indium atom must be *trans*. The details of the structure, however, are still to be established; some extra molecules of pyridine are likely to be found.

Data have been measured on a four-circle diffractometer with X-rays of wavelength 0.71073 Å, to a maximum θ of 25°. Corrections are made for absorption and for a 2% decay in intensities. A total of 5356 measured data yield 3791 unique reflections after averaging of symmetry-equivalents; $R_{int} = 0.022$.

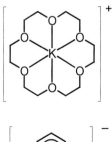

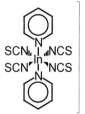

Fig. 3.17 The cation and anion of case study 5.

There are eight inversion centres per unit cell, and one of them can be chosen arbitrarily for one of the metal atoms, corresponding to a choice among the available unit cell origins. It is not possible, however, to place the second metal atom arbitrarily on one of the other inversion centres; they must have the correct relative positions. It is in fact possible to decide on the correct positions of both metal atoms by a careful examination of the distribution of intensities in the data set, but this is not necessary. The indium atom alone has enough electron density to serve as the first model structure. It is placed on the inversion centre at $x = y = z = 0.5$, which sets the anion conveniently at the exact centre of the cell. No Patterson synthesis or direct methods calculation is necessary in order to solve this structure!

The indium atom alone gives $wR2 = 0.7990$ and $R = 0.4602$. Now the largest difference electron density peak lies on another inversion centre and is clearly the potassium atom. The next peaks in descending order of height are two for sulfur, then three for oxygen, four for nitrogen, and 18 for carbon atoms, corresponding to half a cation, half an anion and one separate pyridine molecule in the asymmetric unit (half the unit cell). The assignments are based on peak heights and geometry; in the uncoordinated pyridine molecule, the largest peak is assigned provisionally as nitrogen.

If a nitrogen atom is refined erroneously as a carbon atom, it does not have enough electron density; as a partial compensation, the refinement produces an unusually small displacement parameter, which corresponds to a sharpening of the electron density distribution in an attempt to give a larger peak. Conversely, mis-assignment of a nitrogen atom as an oxygen atom would give a relatively large displacement parameter, spreading out the too high electron density in order to reduce the peak.

Isotropic refinement of these atoms (with the two metal atoms fixed in position on their respective inversion centres) gives $wR2 = 0.2281$, $R = 0.0717$; anisotropic, $wR2 = 0.1375$, $R = 0.0353$. All hydrogen atoms are found in a difference map as the top 22 peaks. The uncoordinated pyridine nitrogen atom is confirmed as correct because it gives the shortest bonds in the ring, the displacement parameters are normal, and it does not have an attached hydrogen atom. Inclusion of hydrogen atoms with constraints gives final values of $wR2 = 0.0529$, $R = 0.0206$, 257 parameters, and no difference map features outside the range ± 0.6 e Å^{-3}. Refinement is complete.

The structures of one cation and one anion are shown in Fig. 3.18. The anion is essentially octahedral with two *trans* pyridine ligands and the thiocyanate ligands are coordinated through nitrogen rather than through sulfur. The cation has a potassium cation coordinated by the macrocyclic crown ether with its six oxygen atoms approximately in one plane. Two pyridine molecules lie above and below the potassium in a weak η^6

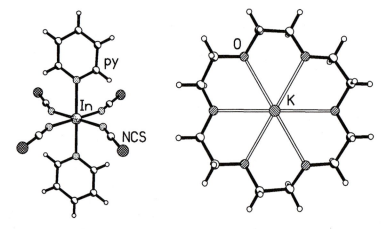

Fig. 3.18   The structures of the cation and anion of case study 5.

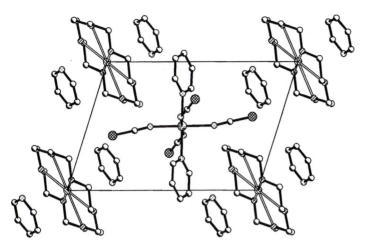

Fig. 3.19 Crystal packing of cations, anions and solvent molecules in case study 5.

coordination; the crystal packing of the cations, anions and weakly coordinated solvent molecules is shown in Fig. 3.19.

This case study illustrates how some structures can be solved from simple symmetry arguments. It provides an example in which the asymmetric unit of the structure is smaller than the chemical formula, with chemical units sitting on crystallographic symmetry elements, and also illustrates the incorporation of extra solvent molecules in the crystal structure.

The symbol η^6 is standard coordination chemistry notation, indicating a ligand attached through six atoms.

3.6   Case study 6: a large polynuclear complex

The complex $[NBu^n_4]_3[(MeO)TiW_5O_{18}]$ is synthesized from a mixture of titanium and tungsten oxo and alkoxo reagents in acetonitrile solution. Data have been collected on an area-detector diffractometer with X-rays of wavelength 0.71073 Å, from a crystal of size $0.34 \times 0.28 \times 0.24$ mm, at 160 K. The crystal is monoclinic, with cell parameters

$$a = 24.283(2) \qquad b = 17.1798(12) \qquad c = 16.5903(12) \text{ Å}$$

$$\beta = 98.095(2)° \qquad [\alpha = \gamma = 90°] \qquad V = 4448.8(6) \text{ Å}^3$$

and the space group is Pc; this is not uniquely determined by the diffraction pattern, and two different space groups (this one non-centrosymmetric, the other centrosymmetric $P2/c$) have to be tried, only one of them successfully giving a structure solution. With $Z = 4$, the average volume per non-hydrogen atom is 21.8 Å^3. This gives no initial information about the molecular structure and does not help in the choice of space group.

A total of 29 642 measured intensities with $\theta < 25.6°$ reduce to 17 063 unique data with $R_{int} = 0.046$ after averaging of symmetry-equivalent reflections; corrections are applied for absorption, but not for intensity decay.

The structure is easily solved by direct methods, giving an electron density map showing all the metal atoms. There are 12 of these in the asymmetric unit, which contains two formula units. The electron densities of titanium and tungsten (atomic numbers 22 and 76) are very different, so they can be distinguished with no difficulty. The remaining non-hydrogen atoms are

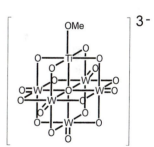

Fig. 3.20   The anion of case study 6.

Space group Pc has a primitive (P) unit cell and a glide plane (with translation in the c direction) perpendicular to the b axis, which is conventionally taken as the main symmetry axis. Space group $P2/c$ additionally has a two-fold rotation axis in this direction (and also inversion symmetry, which is not explicitly given in the space group symbol).

located in subsequent difference maps; because of the size of the structure and the presence of some disorder in the *n*-butyl chains of the cations, several rounds of the calculations (forward and reverse Fourier transforms) are necessary in this case before all the atoms are found. In addition to two anions and six cations, the asymmetric unit is found to contain one molecule of acetonitrile solvent. Following anisotropic refinement, the hydrogen atoms are also located in difference maps and included in refinement with constraints. The disorder of some alkyl chains is modelled by including two alternative positions for each and keeping bond lengths and angles similar for them, by means of *restraints*. Even with such difficulties, final values of 0.0848 for *wR*2 and 0.0350 for *R* are achieved. A total of 1464 parameters are refined. There are some peaks > 1 e Å^{-3} close to metal atoms in a final difference map, but no other features of significant size.

Because the structure is non-centrosymmetric, the absolute structure parameter is refined; with several heavy atoms present, anomalous scattering effects are significant, and a refined parameter of 0.089(11), close to zero, indicates that the structure is correct.

The two anions in the asymmetric unit (Fig. 3.22) have essentially the same geometry except for the torsional position of the OMe group about the Ti—O bond. The anion is the well-known $[W_6O_{19}]^{2-}$ polytungstate ion with one WO group replaced by TiOMe. The cations and solvent are of no particular structural interest and include some disorder.

> Restraints are rather like constraints in principle, in that they provide some control over the refinement process, but they are approximate rather than exact. Mathematically they are treated in quite a different way, which is beyond the scope of this book.

Fig. 3.21   One of the disordered cations of case study 6. Minor components are shown dotted. In each case the disorder consists of alternative orientations for the terminal methyl group of an alkyl chain.

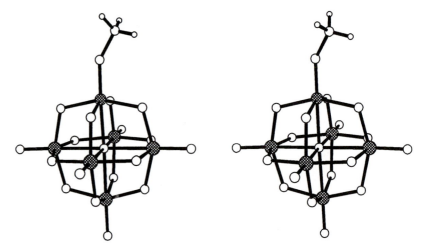

Fig. 3.22   The two crystallographically independent anions in the asymmetric unit of case study 6.

This case study illustrates the advantage of an area-detector system for very large structures, with an overnight data collection which would take many days with a conventional four-circle diffractometer. It provides an example of a structure with more than one chemical formula unit in the structural asymmetric unit, and a partial structure determination by direct methods.

3.7   Case study 7: a microporous material with large channels

This is a synthetic aluminophosphate with a framework structure of basic formula $AlPO_4$, in which a proportion of the aluminium atoms are replaced at random by magnesium. It has large pores containing water molecules and organic cations (heptamethylene diquinuclidinium, two organic cages linked by a chain of seven CH_2 groups) which are necessary for the synthesis. Such materials usually have the guest ions and molecules highly disordered, so that they appear as smeared out electron density with no geometrically interpretable pattern. The formula, from chemical analysis, is $Mg_{0.18}Al_{0.82}$ $PO_4\cdot(cation)_{0.094}\cdot0.22H_2O$.

Only very small crystals of this material are available, essentially a coarse powder. With conventional X-ray sources the diffraction pattern is far too weak to measure. Data were collected with synchrotron radiation, wavelength 0.6956 Å, at 160 K, from a crystal of octahedral shape and 0.04 mm maximum dimension; an area detector was used to record the diffraction pattern in a few hours. Synchrotron radiation shows a significant intensity decay during this time (total about 20%), for which a correction was necessary; a correction was also applied for absorption, although the effects are small for such a tiny crystal.

The structure is tetragonal, space group $P\bar{4}n2$, with unit cell parameters

$$a = b = 13.7745(4) \qquad c = 21.9060(6)\ \text{Å}$$

$$[\alpha = \beta = \gamma = 90°] \qquad V = 4156.4(2)\ \text{Å}^3$$

The correct value for Z is not easy to estimate for such a structure; the '18 Å^3 rule' is not applicable for polymeric inorganic materials. In fact, successful structure determination shows that $Z = 28$ for the approximate formula $AlPO_4$: the unit cell contains 28 metal atoms, 28 phosphorus and 112 oxygen, together with an indeterminate number of organic cations and water molecules. Some of the metal and phosphorus atoms lie on two-fold rotation axes, while the rest and all the oxygen atoms lie in *general positions*. The number of unique framework atoms to be found is four metal, four phosphorus and 14 oxygen.

A total of 26 653 measured intensities with $\theta < 25.9°$ give 4094 unique data, $R_{int} = 0.031$. An automatic direct methods calculation reveals all the framework metal, phosphorus and oxygen atoms. Each metal atom is actually a random mixture of aluminium and magnesium in the ratio 0.82:0.18. It is not easy to distinguish aluminium and magnesium by X-rays in such systems, even when they are ordered on different sites, because the electron density is very similar (atomic numbers 13 and 12), so refinement is carried out as if the sites were purely occupied by aluminium. Isotropic, followed by anisotropic, refinement gives $wR2 = 0.1888$ and $R = 0.0679$ for 192 refined parameters. A difference map shows electron density in the large open pores of the framework, which corresponds to the organic cations and water molecules, but these cannot be resolved as individual localized atoms; only the framework structure can be determined, which is itself an important result, and the unresolved disorder simply makes the overall structure less precise than in many other cases.

This is another non-centrosymmetric structure. The absolute structure parameter refines to 0.5(2), intermediate between the two possibilities and not very precisely determined. The uncertainty is high, because anomalous scattering effects are quite small in the absence of heavy atoms. The structure is likely to be *twinned*, a combination of two inversion-related components in

In this space group symbol, the P stands for primitive unit cell; there is an 'inversion–rotation' four-fold axis ($\bar{4}$, this is called an S_4 axis in the notation used in spectroscopy) rather than a simple four-fold rotation axis along the c axis, diagonal (*n*) glide-planes perpendicular to both a and b axes, and two-fold rotation axes lying between the a and b axes.

Atoms lying on symmetry elements of the space group are said to be in *special positions*, and atoms anywhere else are in *general positions*.

See section 4.1 in the next chapter for a discussion of *twinning*.

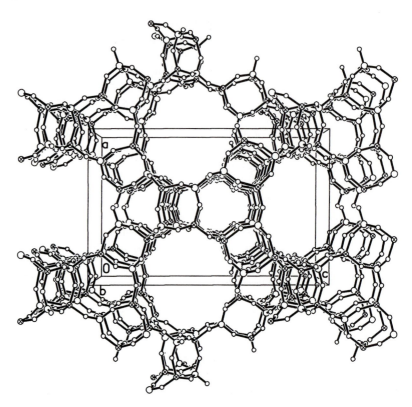

Fig. 3.23   The magnesium-substituted aluminophosphate framework of case study 7.

approximately equal amounts; this effect is not unusual for synthetic materials of this kind.

The framework structure is shown in Fig. 3.23. The large channels, which contain disordered cations and water molecules, are clearly seen.

This final example illustrates a case of extreme disorder, which does not adversely affect the most important results, an incomplete but still very useful structure determination, and the use of synchrotron radiation to give measurable data from very small crystals.

3.8   Problems

The following Patterson solution problems are similar to two of the case studies presented in this chapter. Each of them follows on from one of the problems at the end of Chapter 1, which should be completed first.

Problem 3.1

The triclinic space group for the complex $[(C_{18}H_{18}N_4S)HgBr_2]$ in Problem 1.1 is $P\bar{1}$. For each mercury atom at a position (x, y, z) in the unit cell, space group symmetry requires that there is another mercury atom at the position $(-x, -y, -z)$. Where, apart from the origin $(0, 0, 0)$, will the largest peaks be found in the Patterson map for this structure?

The largest peaks found in the Patterson map calculated from the observed diffraction pattern are listed below (Table 3.5); there are also peaks at symmetry-equivalent positions. All other peaks are under 100 in height.

Deduce the coordinates of one mercury atom in the structure.
To what are peaks 3 and 4, and peaks 5 and 6, probably due?

Table 3.5   The largest Patterson peaks for Problem 3.1

Peak number	x	y	z	Peak height	Vector length (Å)
1	0.000	0.000	0.000	999	0.00
2	0.358	0.374	0.540	336	8.23
3	0.118	−0.124	0.154	224	2.61
4	0.188	0.111	−0.094	223	2.50
5	0.452	0.514	0.558	223	8.46
6	0.471	0.243	0.689	222	6.76

Problem 3.2

The monoclinic space group for the indium complex in Problem 1.2 is $P2_1/c$. For each atom at a general position (x, y, z) in this space group, there must be three symmetry-equivalent atoms at positions $(-x, -y, -z)$, $(-x, \frac{1}{2} + y, \frac{1}{2} - z)$, and $(x, \frac{1}{2} - y, \frac{1}{2} + z)$. Derive from these the positions of the corresponding Patterson vector peaks (similar to Table 3.2, but with different entries).

The four largest peaks in the Patterson map for this compound are at positions given in Table 3.6, together with peaks at symmetry-equivalent positions. Propose (x, y, z) coordinates for one indium atom consistent with these peaks.

Table 3.6   The largest Patterson peaks for Problem 3.2

Peak number	x	y	z	Peak height	Vector length (Å)
1	0.000	0.000	0.000	999	0.00
2	0.000	0.888	0.500	348	8.45
3	−0.120	0.500	0.820	329	9.33
4	−0.120	0.388	0.320	179	8.77

4 Further topics

This chapter deals with a few points which lie outside the main process of crystal structure determination, or which have not been illustrated by the case studies of the previous chapter.

4.1 Disorder and twinning

Case studies 3, 6 and 7 included structural *disorder*. This is a random (not systematic) variation in the detailed contents of the asymmetric unit of a crystal structure. For an ideal structure, all asymmetric units are exactly equivalent under the space group symmetry, and all unit cells are identical. Instantaneously, of course, this is never true, as each atom is undergoing vibration and these movements are not usually correlated throughout the structure, but this effect is dealt with by the atomic displacement parameters, giving equivalence on a time-averaged basis. Large amplitudes of vibration are sometimes referred to as *dynamic disorder*, and they are reduced by cooling the sample in low-temperature data collection.

Static disorder consists of alternative positions for atoms or groups of atoms. If it is truly random, then what X-ray diffraction sees is the average asymmetric unit. This appears in the model structure as partially occupied atom sites.

A commonly observed example is a methyl (CH_3) or trifluoromethyl (CF_3) group attached to an aromatic ring, as in a toluene solvent molecule or a tolyl substituent. There is no single preferred torsional orientation of the methyl group, and the energy barrier to rotation about the C—C bond is relatively low. In some structures, a difference electron density map will show three clear positions for the hydrogen or fluorine atoms, because the CH_3 or CF_3 group is held in one preferred orientation by neighbouring atoms in the same or another molecule. In others, six positions are found (with lower peak heights), corresponding to two alternative orientations with comparable energies, because the intermolecular interactions are weaker, and each molecule in the structure adopts one of the two possibilities at random, with no regard for the orientation adopted in neighbouring asymmetric units (Fig. 4.1).

Other commonly observed disorder patterns are for conformationally flexible groups of atoms, such as long alkyl chains (as in case study 6), non-planar five-membered rings (tetrahydrofuran, thf, is notorious in this respect), counter-ions of high symmetry which are only loosely held in place with no strong intermolecular interactions (particularly BF_4^-, ClO_4^- and PF_6^- anions), and small solvent molecules not anchored by hydrogen bonding or other significant interactions (such as toluene, acetone and dichloromethane). Figure 4.2 shows some typical cases.

Where the disordered atom sites are well resolved, so that individual electron density peaks can be seen, refinement is usually straightforward.

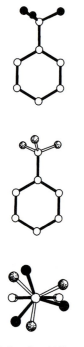

Fig. 4.1 A disordered trifluoromethyl group attached to a benzene ring: top and centre, the two observed disorder components; bottom, the combined disorder model seen in projection along the bond between the CF_3 group (in front) and the ring (behind).

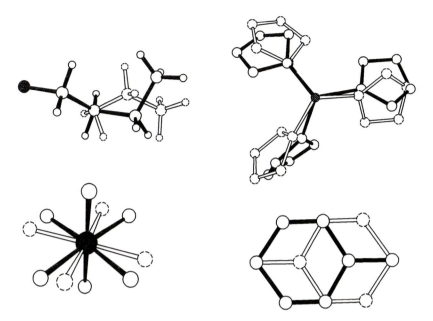

Fig. 4.2 Some examples of disorder: top left, an *n*-butyl chain in which the last two carbon atoms (with their hydrogen atoms) adopt two alternative postions; top right, three disordered thf ligands coordinated to a metal atom; bottom left, an AsF_6^- anion with two orientations related by rotation about one of the linear FAsF units; bottom right, a toluene solvent molecule disordered over an inversion centre (hydrogen atoms not located!).

When disordered atom sites are closer together than normal bonding distances, special techniques are often required, in which the expected molecular geometry is used as data in the refinement alongside the diffraction pattern intensities. These techniques are called *constraints* and *restraints*, and they are particularly important in cases of high disorder.

Examples of *constraints* and *restraints* are given in the previous chapter.

Very often, the disordered part of a crystal structure is not of particular interest, and the fact that this portion is less well determined is not important. Unfortunately, however, disorder in any part of the structure affects the reliability with which the whole structure can be determined. This is because the whole structure generally contributes to the whole diffraction pattern in the experiment, and the whole diffraction pattern contributes to the whole structure in the subsequent calculations; this is the nature of Fourier transforms. The effect can be seen in two particular ways.

First, except in the simplest cases, disorder can be difficult to incorporate in the model structure which is refined, especially when some alternative atom sites lie close together or when there are multiple disorder sites. A less than ideal model structure makes it more difficult to match the calculated and observed diffraction patterns and so leads to higher uncertainties in all the refined parameters than there would be for a fully ordered structure.

Second, static disorder represents a spreading out of electron density from ideal ordered positions, and this, like atomic vibrations, increases interference effects and hence reduces diffracted intensities, particularly at higher Bragg angles (see section 1.6). Badly disordered structures give diffraction patterns in which the intensities fall off rapidly at higher angles; a lower proportion of reflections will be of significant intensity than for an ordered structure of similar scattering power. A shortage of high-angle data with significant intensity leads inevitably to a structure with lower precision, not only because there are fewer data. Inspection of the Bragg equation (1.3 and 1.4) shows that high scattering angle corresponds to small d-spacings, i.e. the *resolution* of closely spaced features in the structure. The maximum Bragg angle for which data are measured dictates the effective minimum resolution to which the structure can be determined. The effect of omitting higher-angle data is illustrated in Fig. 4.3, where electron density maps have been calculated from all data with $\theta < 25°$ ($d > 0.84$ Å) and from the data with $\theta < 15°$ ($d > 1.37$ Å). With the lower resolution data only, it is much more difficult to distinguish the individual lighter atoms.

$$d = \frac{\lambda}{2}\left(\frac{1}{\sin\theta}\right)$$

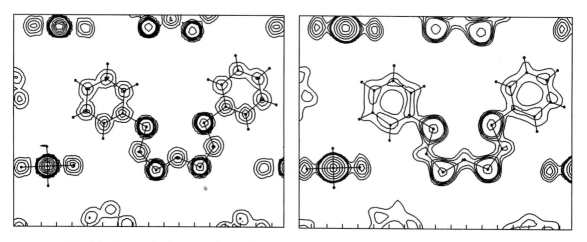

Fig. 4.3   Electron density maps calculated from data with $\theta < 25°$ (left) and from data with $\theta < 15°$ (right).

High disorder particularly affects crystal structures of biological macro-molecules, such as proteins, which incorporate large amounts of disordered solvent water in the substantial spaces between molecules. The diffraction intensities are also weak because of the large size of the molecules. As a consequence, usually only relatively low-angle intensities are observed, and atomic resolution is rarely achieved; in many cases, resolution is limited to 2 Å, 3 Å or even worse. Similar, though less serious, problems can affect some large non-biological structures in developing research areas such as supramolecular chemistry. Low-temperature data collection and the use of synchrotron radiation are both important as means of maximizing intensities and crystal stability.

Another departure from ideal structures which can seriously hinder a crystal structure determination is the phenomenon of *twinning*. A twinned crystal is one in which two (or more) orientations or mirror images of the same structure occur together in a well-defined relationship to each other. It tends to occur when there are fortuitous rational relationships among the unit cell parameters, such as for a monoclinic structure with the angle β close to 90°, or with similar values for the a and c axis lengths. Twinning then results from 'mistakes' in putting the unit cells together to form the complete crystal during its growth, because they can fit almost equally well two different ways round.

A twinned crystal gives a diffraction pattern which is the superposition of the diffraction patterns of the two (or more) individual component parts of the crystal. In some cases, the two diffraction patterns have reflections which coincide exactly, each measured intensity then being the sum of two different but related reflections. In other cases, the presence of two diffraction patterns can be recognized from the outset, because they are not exactly superimposed. Whatever the precise nature of the twin relationship, if it can be worked out from the observed diffraction pattern, then there are methods for solving and refining the structure, though it is more complicated than for a normal untwinned structure.

A common case of twinning which is easily dealt with occurs when a chiral structure grows together with its opposite enantiomer in a single crystal. This can easily happen if both forms of the material are present together in solution, because the unit cell shape is identical for both enantiomers, a crystal lattice always being centrosymmetric even if the overall structure is not (see section 1.3). The reflections which exactly overlap in this case are pairs of h, k, l and $-h, -k, -l$. An attempt to determine the absolute configuration by refinement of a special parameter (section 3.3) will reveal this twinning. For one pure chiral form, the absolute structure parameter should refine to a value of 0 with a small standard uncertainty (s.u.), if the effects of anomalous scattering are significant; if the data actually correspond to the opposite chiral form from that of the model structure, the parameter should be 1 with a small s.u. An intermediate value, around 0.5 with a small s.u., indicates *racemic twinning*. A large s.u. is obtained if the anomalous dispersion effects are not sufficient to provide reliable information on the absolute structure.

More generally, for any non-centrosymmetric structure together with its inverse form; chiral structures are a subset of non-centrosymmetric structures.

4.2   Single-crystal neutron diffraction

X-rays are used for crystal structure determination because they have wavelengths comparable to the separations between atoms in molecules, and so they give measurable diffraction effects from crystals. Any other radiation with a similar wavelength would, in principle, serve the same purpose. Of course, there are no other forms of electromagnetic radiation with the same wavelengths as X-rays, by definition.

According to the de Broglie relationship

$$\lambda = h/p = h/mv \qquad (4.1)$$

an object of mass m moving with velocity v and momentum $p = mv$ has an associated wavelength and can display corresponding wave properties. For neutrons generated by a nuclear reactor, the associated wavelengths lie in the same range as X-rays, so a beam of neutrons can be diffracted by crystalline material.

The use of neutrons for diffraction is experimentally much more difficult and expensive than the use of a conventional laboratory X-ray tube and, in most cases, diffracted intensities are considerably weaker, so there is no point in it unless it offers some significant advantages over X-ray diffraction. For most structure determinations this is not the case, and X-ray diffraction is much more widely used. There are, however, circumstances in which neutrons provide clear advantages, arising from the different ways in which neutrons and X-rays interact with matter as they pass through it.

X-rays, as we have seen, are scattered by the electrons of atoms; an X-ray diffraction experiment shows the electron density distribution within the unit

Fig. 4.4   Total electron density (left) and difference electron density (right) for the location of a hydrogen atom attached to a benzene ring, as obtained from X-ray diffraction at low temperature. The points and lines show the final refined positions of the atoms and bonds, with the C–H bond length extended to its expected internuclear distance (from spectroscopic measurements of many small molecules). The relatively poor scattering and the inward displacement of the hydrogen atom are apparent.

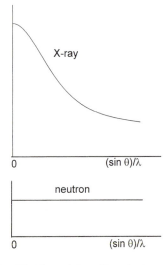

Fig. 4.5   The variation of X-ray (top) and neutron (bottom) atomic scattering factors with Bragg angle for stationary atoms. In both cases atomic vibration causes a reduction in scattering factors at higher Bragg angles.

Table 4.1   Relative X-ray and neutron scattering factors of selected elements and isotopes. The two sets of values are not on the same scale.

Atom	X-ray	Neutron
H	1	−3.7
D	1	6.7
C	6	6.6
N	7	9.4
O	8	5.8
^{35}Cl	17	11.7
^{37}Cl	17	3.1
V	23	−0.4
W	74	4.8
Re	75	9.2
U	92	8.4

cell of a crystal. This electron density distribution is usually interpreted in terms of atomic positions, and leads to molecular geometry. Since the electron density of each atom is generally distributed approximately symmetrically about the nucleus, this interpretation is valid, but in reality there are deviations from spherical symmetry, caused by chemical bonding and other valence effects. The effect is particularly marked for hydrogen atoms, which are consistently located too close to their bonded atoms by X-ray diffraction (Fig. 4.4).

Neutrons, by contrast, interact insignificantly with the electron density when they pass through a crystalline solid. Scattering is instead by the nuclei. Since both nuclei and neutrons are extremely small, significant scattering takes place only when a neutron passes close to a nucleus, and on average the total intensity of diffraction of neutrons by a crystal is low compared with that of X-rays. The relatively weak scattering means that larger crystals are preferred for neutron diffraction, and it may not be easy to grow them. On the other hand, neutron scattering by a stationary atom does not fall off at higher angle like that of X-rays (Fig. 4.5); lower intensities at higher angles are due entirely to atomic vibrations.

Although the scattering power of an atom for X-rays is directly proportional to its atomic number (the number of electrons in the neutral atom), there is no simple relationship between neutron scattering power and atomic number. Neutron scattering powers vary erratically across the periodic table, often with large differences between adjacent elements, and heavier elements do not dominate lighter ones as they do with X-rays; even different isotopes of the same element have different neutron scattering powers. A selection of relative scattering powers for X-rays and neutrons is given in Table 4.1.

It can be seen that some nuclei scatter in phase (positive scattering factors), while others scatter out of phase (negative scattering factors). Note that different isotopes of the same element may have quite different neutron scattering powers; this is particularly so for the isotopes of hydrogen, H and D. Elements (isotopes) with very small scattering powers, such as V, are effectively almost invisible to neutrons. Among common elements, H (D even more so) and N are particularly good neutron scatterers.

There are several important consequences of this difference in the nature of X-ray and neutron scattering, which make neutron diffraction a useful tool in particular cases.

Compared with X-rays, neutrons are generally good at locating light atoms in the presence of much heavier atoms, though this depends very much on the particular elements involved. In particular, the precise location of first-row atoms such as C, N, O in structures containing several very heavy atoms such as W, Re, U is likely to be more successful with neutron diffraction, though an X-ray result is perfectly adequate in most cases (see, for example, case study 6) unless small differences in light-atom bond lengths are to be detected.

An extreme case is, of course, the location of hydrogen atoms, for which neutron diffraction is far superior to X-ray diffraction, especially for deuterated compounds. Not only is the neutron result more *precise*, because H/D atoms scatter relatively strongly, it is also more *accurate*, because it locates the nuclei directly rather than the electron density distorted by valence effects. For studies in which *precise* and *accurate* hydrogen atom location is important, neutron diffraction is the method of choice. Examples include hydride (H^-) ligands in transition metal complexes, bridging hydrogen atoms in electron-deficient compounds such as boranes, and hydrogen bonding. In the majority of structures, however, hydrogen atom positions are entirely predictable and neutron diffraction is not justified.

Neutron diffraction can clearly distinguish many pairs of neighbouring elements in the periodic table, which have almost the same X-ray scattering power. This may be of value in some compounds such as mixed-metal complexes (e.g. containing both W and Re, which have 74 and 75 electrons, respectively, but quite different neutron scattering powers), alloys (where metal atoms may be ordered or disordered) and minerals.

Distinguishing between isotopes of the same element is impossible with X-ray diffraction but, in many cases, straightforward with neutrons, provided the isotopes are not disordered in the structure. A case in point is the determination of the H and D sites in a partially deuterated compound, which may help, for example, in establishing a reaction mechanism by unambiguously identifying the isotopic substitution in the product.

There are more advanced types of experiment, in which both X-rays and neutrons are used to study the same structure. Since neutrons locate nuclei, from which core electron density can be calculated, and X-rays reveal the total electron density distribution, the combination provides a means of mapping valence electrons and bonding effects. Such studies require extremely careful measurements and corrections, since the valence effects are small compared with the total electron density, and they lie beyond the scope of this book.

The distinction between precision and accuracy is important. *Precision* refers to the spread of results obtained if a measurement is repeated many times; it measures repeatability or the degree of confidence with which a particular measurement can be made, and it is measured by statistical parameters such as s.u.'s. *Accuracy* refers to the agreement of the measurement with the true value. Thus, a result can be precise but not accurate (like a wrongly set digital watch), and it can be accurate but not precise.

Under certain circumstances, significant differences between the X-ray scattering factors of neighbouring elements can be generated by choosing a wavelength which gives a large anomalous scattering effect for one of them and a small effect for the other. This requires tuneable X-ray wavelengths, which can be achieved with synchrotron radiation but not with standard laboratory sources.

4.3 Diffraction by powder samples

A single crystal gives a diffraction pattern (with either X-rays or neutrons) with discrete diffracted beams, each in a definite direction relative to the orientation of the crystal and the incident beam, according to the Bragg equation. Because the diffraction conditions are severe, a stationary single crystal gives very few reflections (see section 2.3). In order to generate the complete diffraction pattern it is necessary to rotate the crystal in the X-ray or neutron beam.

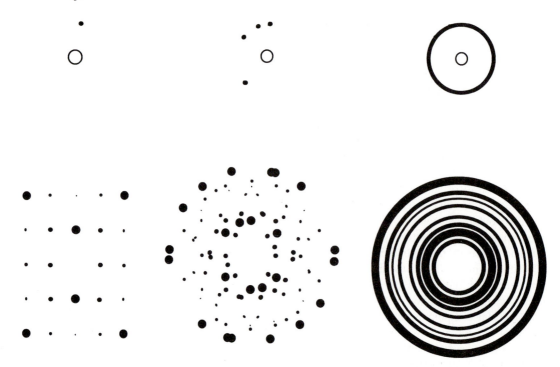

Fig. 4.6   The relationship between single-crystal and powder diffraction. Top: the effect for one individual reflection. The centre of the detector is marked with an open circle. Left: the position of one reflection from a single crystal. Centre: the positions of this reflection derived from four crystals together in different orientations. Right: the effect for a very large number of crystals. Bottom: the effect for a complete simple diffraction pattern. Left: the pattern from one carefully aligned single crystal. Centre: the patterns from four crystals superimposed in random relative orientations. Right: the pattern for a very large number of crystals; this is a powder diffraction pattern, and each spot in the left diagram has generated a complete circle in the right diagram.

Even individual tiny crystals which can be seen only under a microscope, such as constitute a fine powder, are still effectively infinite in size compared with the wavelength of X-rays, so each one acts as a single crystal. In a fine powder, the number of individual crystals is also effectively infinite, with all possible orientations present simultaneously.

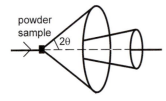

Fig. 4.7   Cones of diffracted X-rays produced by each reflection from a microcrystalline powder sample.

If several single crystals of the same material in different orientations are irradiated simultaneously by X-rays, each of them gives its own diffraction pattern and these are superimposed. As the composite sample is rotated, any particular reflection will be generated by each of the individual crystals at different times as the Bragg equation is satisfied; the Bragg angle and intensity will be the same in each case (assuming equal sizes of crystals), but the direction of the diffracted beam will vary, while always being inclined at 2θ to the straight-through direction (Fig. 4.6). On a flat detector perpendicular to the incident beam and on the opposite side of the sample, this set of corresponding reflections from the multiple crystals appear as identical spots on a circle. With an increasing number of identical and randomly oriented crystals, more such spots appear, all lying on the same circle, which is where a cone of diffracted radiation hits the detector. A microcrystalline powder consists of an essentially infinite number of tiny crystals and this produces a complete circle for a particular reflection.

The same occurs for every Bragg reflection, each one giving a cone of radiation with semi-angle 2θ (Fig. 4.7), and hence producing a circle on the detector. The overall result is a set of many concentric circles, with radii dictated by the Bragg equation and hence the unit cell geometry, and with intensities closely related to those that would be produced by one single crystal.

In practice, a powder diffraction pattern is usually measured either on a strip of photographic film wrapped round the sample in a cylindrical shape (a powder camera), in order to reach high Bragg angles (with θ approaching 90°, the diffracted beam is almost doubled back on the incident beam), or by an electronic detector which is driven in a circle around the sample under computer control (a powder diffractometer). In either case, intensity is recorded as a function of angle, and for each reflection a Bragg angle and an intensity can be obtained. The effect of using a microcrystalline powder instead of a single crystal is to compress the full three-dimensional diffraction pattern into a one-dimensional pattern (the only geometrical variable is θ). It is also possible to use a position-sensitive detector (either an area detector, or one which is one-dimensional rather than two-dimensional in form) to record powder diffraction patterns quickly and efficiently.

For relatively simple structures, giving few reflections, there may be little overlap of these in the powder diffraction pattern. In such cases it is possible to assign individual indices and intensities and carry out structure determination as with single crystals. Even for larger structures, where this is not possible, powder diffraction has important uses, in chemical analysis and for the identification of materials, which is its most common application. A full treatment is given in the Oxford Chemistry Primer on *Inorganic Materials Chemistry* by M. T. Weller. The primary purpose of this brief treatment is to illustrate the relationship between single-crystal and powder diffraction techniques, both of which can be carried out with laboratory X-rays, synchrotron radiation or neutrons.

4.4 Electronic aspects of crystallographic results

A crystal structure determination yields as its results the unit cell geometry and symmetry, and the positions and displacement (vibration) parameters of all the atoms it contains. From these the intramolecular and intermolecular geometry can be calculated, and graphical representations can be produced. All the results, together with the diffraction data from which they are derived, are held electronically in computer files.

The use of computers does not end with the successful completion of a structure refinement. One important further step is the safe storage of the results on computer-readable backup media such as magnetic tapes, for archiving and possible future access.

Computers also play an important role in the publication of structural results in the research literature. Not only are manuscripts prepared with computer word-processors, as in research generally, but there is a growing use of electronic transmission of results from researchers and authors to publishers, via electronic mail and other network services. In this subject, such developments are very much assisted by the well defined nature of the structural results and widespread acceptance of particular standards and formats for them. Recent years have seen the introduction of the so-called *Crystallographic Information File* (CIF), which has been devised as a convenient and flexible form of information for archive, exchange and publication; modern structure refinement programs generally produce a CIF as well as other forms of output, and further items of information can easily be added, since each piece of information in the file is uniquely identified by a name defined in an internationally agreed dictionary. An example of part of a CIF is shown in Fig. 4.8; a complete CIF for a large structure is a long file,

```
_cell_length_a                      17.3546(11)
_cell_length_b                      6.7359(7)
_cell_length_c                      15.7608(9)
_cell_angle_alpha                   90.00
_cell_angle_beta                    90.00
_cell_angle_gamma                   90.00
_cell_volume                        1842.4(2)
_cell_formula_units_Z               4
_cell_measurement_temperature       240(2)
_cell_measurement_reflns_used       32
_cell_measurement_theta_min         10.59
_cell_measurement_theta_max         12.28

_exptl_crystal_description          needle
_exptl_crystal_colour               'very dark blue'
_exptl_crystal_size_max             0.56
_exptl_crystal_size_mid             0.24
_exptl_crystal_size_min             0.16
_exptl_crystal_density_meas         ?
_exptl_crystal_density_diffrn       1.851
_exptl_crystal_density_method       'not measured'
_exptl_crystal_F_000                1016
_exptl_absorpt_coefficient_mu       2.358
_exptl_absorpt_correction_type      empirical
_exptl_absorpt_correction_T_min     0.342
_exptl_absorpt_correction_T_max     0.386

_diffrn_ambient_temperature         240(2)
_diffrn_radiation_wavelength        0.71073
_diffrn_radiation_type              MoK\a
_diffrn_radiation_source            'fine-focus sealed tube'
_diffrn_radiation_monochromator     graphite
_diffrn_measurement_device_type     'Stoe-Siemens diffractometer'
_diffrn_measurement_method
                       '\w/\q scans with on-line profile fitting'
_diffrn_standards_number            3
_diffrn_standards_interval_count    ?
_diffrn_standards_interval_time     60
_diffrn_standards_decay_%           1.7
_diffrn_reflns_number               7487
_diffrn_reflns_av_R_equivalents     0.0506
_diffrn_reflns_av_sigmaI/netI       0.0318
_diffrn_reflns_limit_h_min          -20
_diffrn_reflns_limit_h_max          20
_diffrn_reflns_limit_k_min          -8
_diffrn_reflns_limit_k_max          8
_diffrn_reflns_limit_l_min          -18
_diffrn_reflns_limit_l_max          18
_diffrn_reflns_theta_min            1.75
_diffrn_reflns_theta_max            25.03
_reflns_number_total                1768
_reflns_number_gt                   1444
```

Fig. 4.8   Part of a Crystallographic Information File (CIF), giving experimental details for a crystal structure determination.

including not only the primary results, but also the derived geometry with associated s.u.'s. The diffraction data can also be stored in a defined CIF format, as can an entire research report or manuscript for publication. Computer programs are then used for converting information from a CIF into text and tables more suitable for human readers, by-passing any need for manual typesetting with its inherent probability of introducing errors. Crystallography is very well suited to electronic publishing.

Once structural results have been published in the primary research literature (scientific journals), they are available for anyone to access and use. It is, however, not necessary to work through libraries of printed material to find results of relevance and interest, because of the availability of *computer databases*. Databases are essentially collections of items of information with a common structure and format. Their advantages over paper-based or other storage and retrieval systems include their ease of maintenance and updating, the possibility of automatic validation of new entries, facilities for selecting and sorting entries, and computer-based analysis of selected entries. A database has two components: the stored contents, and suitable software for search, retrieval and analysis.

Computer databases are important in many areas of chemistry and other sciences, and cover such aspects as bibliography and literature citations, safety information, spectroscopy and reaction mechanisms. They are particularly well suited for crystal structures.

Four main structural databases are used internationally in research. The *Metals Crystallographic Data File* (MCDF), administered by the National Research Council of Canada, holds information on metals, alloys and some binary compounds (hydrides and oxides) of metals. In 1997 there were over 55 000 entries.

The *Inorganic Crystal Structure Database* (ICSD) is managed by the Gmelin Institute and the Fachinformationszentrum (Karlsruhe) in Germany. It contains inorganics and minerals, in which there is no organic carbon. In 1997 there were over 43 000 entries.

The *Protein Data Bank* (PDB), maintained by the Brookhaven National Laboratory in New York, USA, stores data for proteins and nucleic acids. A relative newcomer, its size has grown enormously, with over 6000 entries in 1997.

The largest structural database is the *Cambridge Structural Database* (CSD), developed by the Cambridge Crystallographic Data Centre, UK. Its contents are organics, organometallics and metal complexes, and numbered around 168 000 in 1997, with continued rapid expansion.

A further database of interest is the *Crystal Data Identification File* from the US National Institute of Standards and Technology. This holds details of unit cell parameters for over 210 000 structures. It is particularly useful in the early stages of crystal structure determination, when the unit cell geometry has been obtained. If this proves to be essentially the same as an entry in the database, and the chemical composition for that entry is reasonable for the sample being investigated, a unnecessary full data collection and repeat determination of a known structure can be avoided.

Each of these databases has its own individual special features appropriate to the contents, but they also have common aspects. Since the CSD is the most widely used, we consider it further here in illustrating some points.

Entries for the CSD are drawn mainly from the primary research literature. Many of these require manual keyboard input from paper copies, though image scanners can also be used. Some journals request or require electronic submission of structural results, which are then forwarded to the CSD compilers, for example in CIF form. Other entries are supplied direct by crystallographers for inclusion in CSD, whether paper-based publication occurs or not. Individual entries are thoroughly checked for consistency and possible errors, which are either corrected or flagged.

Each entry in the database contains bibliographic information; a collection of individual text and numeric data such as unit cell parameters, temperature of data collection, and *R* factor; a two-dimensional representation of the chemical structural formula; and all the atom positions, from which detailed geometry can be calculated. Searches can be made through the contents against any of these items; particularly useful is the facility to search for all structures containing a particular group of atoms (a molecular fragment), with or without specific restrictions on its geometry. The possible output includes display of all the searchable items, a three-dimensional graphical representation of the structure which can be manipulated interactively, and statistical analysis of any of the numerical items, including specific geometrical features such as bond lengths.

The structural databases are thus an invaluable resource of reliable information, far more convenient to use than the original published literature. They can be used to find a particular structure for various reasons (this includes avoiding repeating work which has already been done!), to obtain information on a series of related structures, to generate a geometry for a structural fragment for use in other calculations such as molecular orbital theory or molecular modelling, and for extensive research into trends and patterns in structures (such as conformations of rings, hydrogen bonding, intermolecular interactions, substituent effects, etc.).

There is no doubt that the science of crystal structure determination, at its best, is a highly developed, well organized, very reliable and powerful technique, from the initial measurement of diffraction data, through the solution of the 'phase problem' and the refinement of trial structures, to the presentation and interpretation of results and their availability in convenient form to the world at large. Much of modern chemistry depends heavily on it, and it continues to be a subject of rapid development and excitement.

There is, of course, much that the databases do *not* contain, such as the authors' discussion of their results, for which the original literature must be consulted; but even here, the databases provide the necessary bibliographic information as a way into the literature.

4.5 Problems

Problem 4.1

What advantages would there be in the use of neutron diffraction for crystal structure determination of each of the following with one exception, and why would X-ray diffraction be preferable for that one case?
(a) The product of a reaction of an organic compound with D_2O in a study of stereochemistry.
(b) A polynuclear osmium carbonyl complex in which differences in the C—O bond lengths of terminal and bridging ligands is of interest.
(c) A natural product containing C, H, N, and O for which the chemical identity needs to be confirmed.
(d) An aluminosilicate mineral which may have the framework Al and Si atoms ordered or disordered.
(e) A platinum complex of a boron hydride which may involve Pt—H—B bridging bonds.

Appendix

Answers to problems

Problem 1.1

Using the method of section 1.5, the unit cell mass is the product of density and cell volume; multiplying by Avogadro's number gives a unit cell mass of 1448.6 daltons. The formula mass is 682.8, and the ratio of these two numbers is 2.12. Thus, there are two molecules of complex per unit cell.

Unit cell mass
$$= 2.16 \times 1113.5 \times (10^{-8})^3$$
$$\times 6.023 \times 10^{23} = 1448.6 \text{ daltons}$$
$$Z = 1448.6/682.8 = 2.12$$

The true formula mass is thus half the unit cell mass, 724.3 daltons. This is greater than the molecular mass of the complex itself by 41.5 daltons. The molecular mass of acetonitrile is 41.1, so there is one molecule of solvent per molecule of complex, or two molecules of solvent per unit cell.

True formula mass $= 1448.6/2 = 724.3$
$724.3 - 682.8 = 41.5$

Problem 1.2

The unit cell volume, calculated as for the monoclinic example in section 1.5, is 4276.4 Å^3. The unit cell mass (product of density and cell volume) is 3605.9 daltons. The ratio of this and the formula mass of 542.4 is 6.65. This is a long way from any of the expected values. The correct value cannot be greater than 6.65, as this would correspond to a negative discrepancy! Presumably the correct answer is 4, and the large discrepancy means a high proportion of thf molecules. Obviously there are equal numbers of cations and anions in the unit cell for charge balance: four of each.

$V = abc \sin\beta = 4276.4 \text{ Å}^3$
Unit cell mass
$$= 1.40 \times 4276.4 \times (10^{-8})^3$$
$$\times 6.023 \times 10^{23} = 3605.9 \text{ daltons}$$
$$Z = 3605.9/542.4 = 6.65$$

The true total formula mass is thus one-quarter of the cell mass, 901.5 daltons. This is 359.1 daltons greater than the formula mass originally assumed, giving five molecules of thf (molecular mass 72.1) per cation–anion pair.

True formula mass $= 3605.9/4 = 901.5$
$901.5 - 542.4 = 359.1$
$359.1/72.1 = 4.98$

In fact, in the fully determined structure, the five thf molecules are all coordinated to the potassium cation (which also interacts significantly with one of the thiocyanate ligands of an adjacent anion, to give six-coordination).

Problem 3.1

With just two symmetry-equivalent positions in the unit cell, the corresponding Patterson peaks will be the differences between them: (position 1) − (position 2), and (position 2) − (position 1). For the positions as given, these differences are $(2x, 2y, 2z)$ and $(-2x, -2y, -2z)$, exactly as shown in case study 1.

$(x, y, z) - (-x, -y, -z) = (2x, 2y, 2z)$
$(-x, -y, -z) - (x, y, z) = (-2x, -2y, -2z)$

The largest peak in the Patterson map, apart from the origin, is peak 2 in Table 3.5. If this is assumed to be the $(2x, 2y, 2z)$ vector between two symmetry-related mercury atoms in the unit cell, then the mercury atom coordinates are obtained simply by dividing by 2: $x = 0.179$, $y = 0.187$, $z = 0.270$.

$2x = 0.358, 2y = 0.374, 2z = 0.540$

After mercury, the largest atoms (most electrons) are bromine, so the next largest Patterson peaks should correspond to vectors between mercury and bromine atoms. Peaks 3 to 6 are all of similar height. Two of them (3 and 4) have vector lengths appropriate for Hg—Br bonds and so are intramolecular vectors corresponding to the two bonds within a molecule. The other two have much longer vector lengths, and so are intermolecular vectors from mercury in one molecule to the bromine atoms in the other molecule.

This problem is very similar to case study 1.

Problem 3.2

A table of vectors can be constructed just like that for case study 2. In fact, it will be the same as Table 3.2, except that all entries $\frac{1}{2} + x$ are replaced by x, $\frac{1}{2} - x$ is replaced by $-x$, and $\frac{1}{2}$ is replaced by 0 where it occurs as the first of the three coordinates in each case. Thus, for example, the four entries in the first column of the table should be $(0, 0, 0)$, $(2x, 2y, 2z)$, $(2x, \frac{1}{2}, \frac{1}{2} + 2z)$ and $(0, \frac{1}{2} + 2y, \frac{1}{2})$.

$2x = -0.120$ (peaks 3 and 4)
$2y = 0.388$ (peak 4)
$\frac{1}{2} + 2y = 0.888$ (peak 2)
$2z = 0.320$ (peak 4)
$\frac{1}{2} + 2z = 0.820$ (peak 3)

The first of these entries is the origin peak (number 1 in Table 3.6). The second is peak 4, the third is peak 3, and the fourth is peak 2. Solving the very simple equations for the coordinates in each case, and checking that they are all consistent with each other, $x = -0.060$, $y = 0.194$, and $z = 0.160$. Note that there are other possible solutions, all of which are equally valid, obtained by using different columns or rows of the full 4×4 vector table. They differ from the solution given here by various combinations of adding the above coordinates to, or subtracting them from, any integer or half-integer (for example, y could equally well be given as -0.194, 0.694, or 0.306, among other values). These different, but equivalent, solutions simply correspond to different possible choices of origin for the unit cell.

Problem 4.1

(a) Neutrons, unlike X-rays, can distinguish between the isotopes H and D, and so provide information on the positions of deuteriation.

(b) Neutrons provide more precise positions for the carbon and oxygen atoms, which are relatively poorly determined with X-rays in the presence of the very heavy (electron-rich) osmium atoms.

(c) There is no particular need for neutron diffraction in this case; X-ray diffraction is perfectly adequate, much less expensive, and more accessible.

(d) Aluminium and silicon have very similar X-ray scattering, so are not easily distinguished, but there is a clear difference in their neutron scattering, which helps resolve the question of order/disorder.

(e) Hydrogen atoms are not easily seen by X-rays in the presence of platinum, especially if they are close to the platinum atom. For neutrons, hydrogen is a strong scatterer and is much more easily, precisely and accurately located. Replacement of H by D would be even better.

Bibliography

Sources of examples

Most of the examples used in this book are drawn from the work of my own research group, though a few have been taken from the published literature with the aid of the Cambridge Structural Database described in Chapter 4 and in the following publication.

1. F. H. Allen and O. Kennard. *Chemical Design and Automation News* 1993, **8**, 1 and 31–37.

References are given below for the major examples in the text.

2. C. M. Aherne, A. J. Banister, I. Lavender, S. E. Lawrence, J. M. Rawson and W. Clegg. *Polyhedron* 1996, **15**, 1877–1886 (example of structure completion).
3. T. Alsina, W. Clegg, K. A. Fraser and J. Sola. *Journal of the Chemical Society, Dalton Transactions* 1992, 1393–1399 (case study 1).
4. C. J. Carmalt, W. Clegg, M. R. J. Elsegood, B. O. Kneisel and N. C. Norman. *Acta Crystallographica, Section C* 1995, **51**, 1254–1258 (case study 5 and symmetry example 3).
5. R. J. Cernik, W. Clegg, C. R. A. Catlow, G. Bushnell-Wye, J. V. Flaherty, G. N. Greaves, I. Burrows, D. J. Taylor, S. J. Teat and M. Hamichi. *Journal of Synchrotron Radiation* 1997, **4**, 279–286 (case study 7).
6. W. Clegg, M. R. J. Elsegood, R. J. Errington and J. Havelock. *Journal of the Chemical Society, Dalton Transactions* 1996, 681–690 (case study 6).
7. W. Clegg, D. A. Greenhalgh and B. P. Straughan. *Journal of the Chemical Society, Dalton Transactions* 1975, 2591–2593 (symmetry example 1).
8. S. Abrahamsson and M. M. Harding. *Acta Crystallographica* 1966, **20**, 377–383 (example of structure solution).

The following have not yet been published.

9. W. Clegg, K. A. Fraser, R. J. Errington *et al.* (symmetry example 2).
10. W. Clegg, L. Horsburgh, R. E. Mulvey *et al.* (case studies 2 and 3).
11. W. Clegg, K. Wade *et al.* (case study 4).

Further reading

The following books (listed in alphabetical order of first author) provide more detailed accounts of the subject.

X-ray analysis and the structure of organic molecules. J. D. Dunitz (1995). Wiley-VCH, Weinheim.

Fundamentals of crystallography. C. Giacovazzo, H. L. Monaco, D. Viterbo, F. Scordari, G. Gilli, G. Zanotti and M. Catti (1992). Oxford University Press.

Crystal structure analysis for chemists and biologists. J. P. Glusker, M. Lewis and M. Rossi (1994). Wiley-VCH, Weinheim.

Crystal structure analysis: a primer. 2nd edition. J. P. Glusker and K. N. Trueblood (1985). Oxford University Press.

The basics of crystallography and diffraction. C. Hammond (1997). Oxford University Press.

Structure determination by X-ray crystallography. 3rd edition. M. F. C. Ladd and R. A. Palmer (1993). Plenum, New York.

Modern X-ray analysis on single crystals. P. Luger (1980). De Gruyter, Berlin.

Essentials of crystallography. D. McKie and C. McKie (1986). Blackwell, Oxford.

Crystallography. D. Schwarzenbach (1996). Wiley, Chichester.

X-ray structure determination. A practical guide. 2nd edition. G. H. Stout and L. H. Jensen (1989). Wiley, New York.

Inorganic materials chemistry. M. T. Weller (1994). Oxford University Press.

An introduction to X-ray crystallography. 2nd edition. M. M. Woolfson (1997). Cambridge University Press.

Index

1375